High Performance Textiles

High Performance Textiles

Editors

Tarek M. Abou Elmaaty
Maria Rosaria Plutino

MDPI • Basel • Beijing • Wuhan • Barcelona • Belgrade • Manchester • Tokyo • Cluj • Tianjin

Editors
Tarek M. Abou Elmaaty
Galala University
Egypt

Maria Rosaria Plutino
Istituto per lo Studio dei Materiali Nanostrutturati
Consiglio Nazionale delle Ricerche (ISMN-CNR)
Italy

Editorial Office
MDPI
St. Alban-Anlage 66
4052 Basel, Switzerland

This is a reprint of articles from the Special Issue published online in the open access journal *Polymers* (ISSN 2073-4360) (available at: https://www.mdpi.com/journal/polymers/special_issues/High_Perform_Text).

For citation purposes, cite each article independently as indicated on the article page online and as indicated below:

LastName, A.A.; LastName, B.B.; LastName, C.C. Article Title. *Journal Name* **Year**, *Volume Number*, Page Range.

ISBN 978-3-0365-3851-8 (Hbk)
ISBN 978-3-0365-3852-5 (PDF)

Cover image courtesy of Maria Rosaria Plutino.

© 2022 by the authors. Articles in this book are Open Access and distributed under the Creative Commons Attribution (CC BY) license, which allows users to download, copy and build upon published articles, as long as the author and publisher are properly credited, which ensures maximum dissemination and a wider impact of our publications.

The book as a whole is distributed by MDPI under the terms and conditions of the Creative Commons license CC BY-NC-ND.

Contents

About the Editors . . . . **vii**

Tarek AbouElmaaty, Shereen A. Abdeldayem, Shaimaa M. Ramadan, Khaled Sayed-Ahmed and Maria Rosaria Plutino
Coloration and Multi-Functionalization of Polypropylene Fabrics with Selenium Nanoparticles
Reprinted from: *Polymers* **2021**, *13*, 2483, doi:10.3390/polym13152483 . . . . **1**

Li-Hua Lyu, Wen-Di Liu and Bao-Zhong Sun
Electromagnetic Wave-Absorbing and Bending Properties of Three-Dimensional Honeycomb Woven Composites
Reprinted from: *Polymers* **2021**, *13*, 1485, doi:10.3390/polym13091485 . . . . **15**

Mary Morris, Xiaofei Philip Ye and Christopher J. Doona
Dyeing Para-Aramid Textiles Pretreated with Soybean Oil and Nonthermal Plasma Using Cationic Dye
Reprinted from: *Polymers* **2021**, *13*, 1492, doi:10.3390/polym13091492 . . . . **31**

Kabir Hossain, Thennarasan Sabapathy, Muzammil Jusoh, Mahmoud A. Abdelghany, Ping Jack Soh, Mohamed Nasrun Osman, Mohd Najib Mohd Yasin, Hasliza A. Rahim and Samir Salem Al-Bawri
A Negative Index Nonagonal CSRR Metamaterial-Based Compact Flexible Planar Monopole Antenna for Ultrawideband Applications Using Viscose-Wool Felt
Reprinted from: *Polymers* **2021**, *13*, 2819, doi:10.3390/polym13162819 . . . . **49**

Tarek Abou Elmaaty, Hanan G. Elsisi, Ghada M. Elsayad, Hagar H. Elhadad, Khaled Sayed-Ahmed and Maria Rosaria Plutino
Fabrication of New Multifunctional Cotton/Lycra Composites Protective Textiles through Deposition of Nano Silica Coating
Reprinted from: *Polymers* **2021**, *13*, 2888, doi:10.3390/polym13172888 . . . . **67**

Sarianna Palola, Farzin Javanshour, Shadi Kolahgar Azari, Vasileios Koutsos and Essi Sarlin
One Surface Treatment, Multiple Possibilities: Broadening the Use-Potential of Para-Aramid Fibers with Mechanical Adhesion
Reprinted from: *Polymers* **2021**, *13*, 3114, doi:10.3390/polym13183114 . . . . **79**

Ana María Rodes-Carbonell, Josué Ferri, Eduardo Garcia-Breijo, Ignacio Montava and Eva Bou-Belda
Influence of Structure and Composition of Woven Fabrics on the Conductivity of Flexography Printed Electronics
Reprinted from: *Polymers* **2021**, *13*, 3165, doi:10.3390/ polym13183165 . . . . **91**

Hany El-Hamshary, Mehrez E. El-Naggar, Tawfik A. Khattab and Ayman El-Faham
Preparation of Multifunctional Plasma Cured Cellulose Fibers Coated with Photo-Induced Nanocomposite toward Self-Cleaning and Antibacterial Textiles
Reprinted from: *Polymers* **2021**, *13*, 3664, doi:10.3390/polym13213664 . . . . **109**

Tarek Abou Elmaaty, Sally Raouf, Khaled Sayed-Ahmed and Maria Rosaria Plutino
Multifunctional Dyeing of Wool Fabrics Using Selenium Nanoparticles
Reprinted from: *Polymers* **2022**, *14*, 191, doi:10.3390/polym14010191 . . . . **121**

Caleb Metzcar, Xiaofei Philip Ye, Toni Wang and Christopher J. Doona
Soybean Oil-Based Biopolymers Induced by Nonthermal Plasma to Enhance the Dyeing of Para-Aramids with a Cationic Dye
Reprinted from: *Polymers* **2022**, *14*, 628, doi:10.3390/polym14030628 . . . . **137**

Tarek Abou Elmaaty, Khaled Sayed-Ahmed, Radwan Mohamed Ali, Kholoud El-Khodary and Shereen A. Abdeldayem
Simultaneous Sonochemical Coloration and Antibacterial Functionalization of Leather with Selenium Nanoparticles (SeNPs)
Reprinted from: *Polymers* **2022**, *14*, 74, doi:10.3390/polym14010074 . . . . **157**

About the Editors

Tarek M. Abou Elmaatyz

Tarek M. Abou Elmaaty is Galala University's first Vice President for academic affairs. He co-supervised in establishing Galala University within his role as a member of the higher committee of establishing national universities, MOHE. He's also a Professor of Textile Dyeing, Damietta University. Professor Tarek Abou El-Maaty earned a Bachelor's degree in Science as well as a Masters and a Doctorate in Organic Chemistry. He started his academic career as a lecturer in Mansoura University in 1998 before being appointed Dean of the Faculty of Applied Arts, Damietta University. He served as a member of the scientific committee for promotion of professors of Applied Art in the Supreme Council of Universities before becoming the Vice President of Damietta University for academic affairs. Prof. Abou El-Maaty's research projects revolve around three themes; heterocyclic chemistry, the development of new dyes, and the modification of fabrics and colorant to produce multifunctional textiles. During the past 15 years, he worked on the synthesis of new functional dyes as well as the modification of fabrics to produce multifunctional textiles to be introduced to the market. He succeeded in creating several new dyes and modified fabrics which have antibacterial activity and multifunctional properties. And currently, his group is specialized in dyeing & finishing of textiles under supercritical carbon dioxide medium and this is the only group in the Middle East working with this technology. Also, they established the international joint China-Egypt supercritical dyeing in 2018. Prof. Tarek Abou El-Maaty received a number of awards including the best researcher Award, Publication Award and recently he won the gold medal (textile field) in Cairo Innovates for two consecutive cycles in 2018 & 2019. He has three patents in the field of textile chemical technology registered at ASRT. He also has over 60 publications and 5 books.

Maria Rosaria Plutino

Maria Rosaria Plutino gas received her Ph.D. in Chemical Sciences at the Univ. of Messina (IT) in 1997. In March 1996 she won a 24-month post-doctoral grant (Human Capital and Mobility), funded by the European Community at the Dep. Oorganisk1 of the Lund Chemical Center (Sweden). In 1999 she had a temporary research contract at the Institute of Chemistry and Technology of Natural Products (ICTPN-CNR), Messina. Since 2001 she has a permanent researcher position at the Institute for the Study of Nanostructured Materials of the CNR (ISMN) in Palermo (c/o Dip ChiBioFarAm, Univ. of Messina). Dr. Plutino is also co-founder, President, and Scientific Manager of the recently constituted ATHENA Green Solutions S.r.l., an Innovative Start-Up, and Joint and not attended Spin-off by the National Research Council (CNR) and by the Univ. of Messina. The basis of the Innovative Start-up is the Arginare entrepreneurial/patent idea, which has participated in various local, regional, and national business competitions, achieving excellent placements and receiving awards and prizes. The company operates in the field of Environmental Technologies, offering in addition to the materials/prototypes of ArgiNaRe, other products, methodologies, consultancy and services in multidisciplinary sectors and the perspective of innovation, sustainability, and recycling. In particular, the mission and objectives of the company ATHENA Green Solutions are to research and production systems and/or prototypes for the resolution of problems deriving from high environmental impact activities with particular reference to seagoing vessels and marine/coastal, industrial and urban pollution. Her main research interest is the rational design, development, and structural study of organometallic complexes of transition metals, supramolecular systems, and nanostructured functional hybrid materials, for applications in the field of optoelectronics, sensors, catalysis, environmental remediation, and biomedicine. She has acquired specific skills in the use of different techniques, such as UV-Vis and Fluorescence Spectrophotometry, NMR (homo- and heteronuclear, mono- and two-dimensional) and FT-IR spectroscopy, Mechanical Calculations and organometallic syntheses in anaerobic and high empty. In particular, the research activity of Dr. Plutino is focused on the development of new innovative and advanced, multi-component and multifunctional, nano-hybrid systems (nanohybrids and nanocomposites), obtained thanks to the use of sol-gel and polymerization techniques carried out in the presence of organic/inorganic hybrid silanes or polymeric precursors and organic and inorganic functional nanofillers, which show implemented physical and surface chemical and physical properties, and which can present potential applications in different sectors such as building, naval, textile, environmental, cultural heritage, biomedical, sensoristic, catalytic. Recently, Dr. Plutino has set-up green and eco-friendly synthesis protocols starting from natural substances or waste, which lead to the obtainment of functional materials that can also be recycled and re-used. The research activity of Dr. Plutino is documented by more than 50 papers published in journals with international circulation and two patents in the evaluation phase. She is also a guest-editor and reviewer of different international scientific journals. Part of the results obtained are the outcome of numerous national and international multidisciplinary scientific collaborations, and are included in various research financed projects ("Blue Growth" and "Cultural Heritage" theme lines in the PO FESR 2014–2020 and PON 2017–2020, where she is the scientific CNR coordinator); they have also been presented at national and international congresses in the form of oral communications and posters.

Article

Coloration and Multi-Functionalization of Polypropylene Fabrics with Selenium Nanoparticles

Tarek AbouElmaaty [1,*], Shereen A. Abdeldayem [2], Shaimaa M. Ramadan [2], Khaled Sayed-Ahmed [3] and Maria Rosaria Plutino [4]

1 Department of Material Art, Galala University, Galala 43713, Egypt
2 Department of Textile Printing, Dyeing and Finishing, Faculty of Applied Arts, Damietta University, Damietta 34512, Egypt; shereen.abdeldayem.82@gmail.com (S.A.A.); designer_shemo@yahoo.com (S.M.R.)
3 Department of Agricultural Chemistry, Faculty of Agriculture, Damietta University, Damietta 34512, Egypt; dr_khaled@yahoo.com
4 Istituto per lo Studio dei Materiali Nanostrutturati, Consiglio Nazionale delle Ricerche, Vill. S. Agata, 98166 Messina, Italy; plutino@pa.ismn.cnr.it
* Correspondence: tasaid@gu.edu.eg

Abstract: In this study, we developed a new approach for depositing selenium nanoparticles (SeNPs) into polypropylene (PP) fabrics via a one-step process under hydrothermal conditions by using an IR-dyeing machine to incorporate several functionalities, mainly coloration, antibacterial activity and ultraviolet (UV) protection. The formation, size distribution, and dispersion of the SeNPs were determined using X-ray diffraction (XRD), ultraviolet-visible (UV/Vis), transmission electron microscopy (TEM) and the color strength, fastness, antibacterial properties, and UV protection of the treated fabrics were also explored. The UV-Vis spectra and TEM analysis confirmed the synthesis of spherical well-dispersed SeNPs and the XRD analysis showed the successful deposition of SeNPs into PP fabrics. The obtained results demonstrate that the SeNPs-PP fabrics is accompanied by a noticeable enhancement in measurements of color strength, fastness, and UV-protection factor (UPF), as well as excellent antibacterial activity. Viability studies showed that SeNPs-PP fabrics are non-toxic against wi-38cell line. In addition, the treated SeNPs-PP fabrics showed an increase in conductivity. The obtained multifunctional fabrics are promising for many industrial applications such as the new generation of curtains, medical fabrics, and even automotive interior parts.

Keywords: selenium nanoparticles; polypropylene; coloration; antibacterial; conductivity; UV protection

Citation: AbouElmaaty, T.; Abdeldayem, S.A.; Ramadan, S.M.; Sayed-Ahmed, K.; Plutino, M.R. Coloration and Multi-Functionalization of Polypropylene Fabrics with Selenium Nanoparticles. *Polymers* **2021**, *13*, 2483. https://doi.org/10.3390/polym13152483

Academic Editor: Andrea Zille

Received: 7 July 2021
Accepted: 23 July 2021
Published: 28 July 2021

Publisher's Note: MDPI stays neutral with regard to jurisdictional claims in published maps and institutional affiliations.

Copyright: © 2021 by the authors. Licensee MDPI, Basel, Switzerland. This article is an open access article distributed under the terms and conditions of the Creative Commons Attribution (CC BY) license (https://creativecommons.org/licenses/by/4.0/).

1. Introduction

Numerous washable or disposable healthcare and hygiene textile products are used in hospitals either for the protection of staff and patients (drapes, beddings, masks, uniforms, wound dressings, bandaging materials, etc.) [1]. Due to the large surface areas, textiles have superior abilities to retain warmth, moisture, and nutrients from spillages and exudates, making them ideal substrates for microorganisms to grow on [2]. Some studies have suggested that healthcare textiles can act as reservoirs and vehicles for the spread of microorganisms in hospitals [3]. With the development of nanotechnology, some inorganic nanoparticles (NPs), such as silver, copper, and zinc, have been identified as promising candidates in combating pathogenic microorganisms [4–9]. Nanotechnology driven therapies with metal and metal oxide nanoparticles (NPs) are emerging as a promising alternative to antibiotics. The high reactivity of these NPs, due to large surface-to-volume ratio, results in intrinsic targeted antimicrobial efficiency even when they are applied in small amounts [10]. Proven activity of metal and metal oxide NPs against wide range of microorganisms including bacteria, fungi, viruses, and other eukaryotic microorganisms [11] was inspiring for researchers to immobilize them onto textiles [12–15]. Selenium nanoparticles (SeNPs) has become the new research target because they are found to possess excellent bioavailability, low toxicity, and contribute to a wide spectrum of health promotion, as well as disease prevention and treatment activities [16]. Biswas et al. [17] prepared silver or selenium nanoparticles on polymeric scaffolds and compared the cytotoxicity of the scaffolds towards mouse fibroblasts using an indirect contact method; the results indicated that the Ag-loaded scaffolds showed high cytotoxicity, while the Se-loaded scaffolds were not toxic to the cells. The low cytotoxicity of SeNPs indicates the great potential for use in biomedical applications [3]. The research on SeNPs as antimicrobial agents is still limited. Several studies have pointed out the ability of SeNPs to exhibit anticancer [18], antioxidant [19], antibacterial and anti-biofilm [20] properties. So far, remarkable antimicrobial activity of these nanoparticles has been evidenced against pathogenic bacteria, fungi, and yeasts [21–23], which inspired researchers to immobilize them onto textiles.

Polypropylene (PP) fabric has excellent physical and mechanical properties [24]. It is a hydrophobic fabric, and several surface modification techniques are adopted to improve wetting, adhesion to polymer surfaces by introduction of a variety of polar groups [25–27]. PP advantages include a great supply, good process, low energy demand, low cost and high chemical stability [28]. There is a great demand for antibacterial PP fabric to be used in different medical applications [29,30]. To our knowledge, there are no reports dealing with coloring PP by using SeNPs without any additives and studying the antibacterial and UV-protection properties, cytotoxicity, and electrical conductivity of the resulted SeNPs-PP fabrics. In this study, SeNPs are prepared in one step process under a simple redox system based on the method mentioned by Abou Elmaaty et al. [31] followed by application of the SeNPs to PP fabric in one-step process under hydrothermal conditions. The characteristics of SeNPs in solution phase were studied by UV-visible spectrophotometer (UV/Vis), transmission electron microscopy (TEM) and X-ray diffraction (XRD). Moreover, scanning electron microscopy (SEM), colour characteristics and antibacterial and UV-protection properties, cytotoxicity and electrical conductivity of the SeNPs-PP fabrics were also evaluated. In summary, we developed a simple, green, and feasible route to produce green coloration of PP-based fabric for multifunctional applications. The low cytotoxicity of SeNPs and antibacterial properties indicates their great potential for use in biomedical applications.

2. Materials and Methods

2.1. Fabric and Chemicals

Polypropylene fabric (100%) was supplied by Shikisen-sha Company (Osaka, Japan) with crystallinity (50.6%), melting enthalpy (105.8 J/g) [32], density (0.91 g/c.c), moisture regain (0%) and tenacity (3.5–8.0 g/den). Sodium hydrogen selenite, ascorbic acid and polyvinylpyrrolidone (PVP) were purchased from *LobaChemie, India*. Other chemicals were commercial grade.

2.2. Green Synthesis of Selenium Nanoparticles (SeNPs)

SeNPs were synthesized via a redox reaction based on the method reported by AbouElmaaty al. with an improved modification [31]. PVP (6g) was dissolved in 100 mL of sodium hydrogen selenite solution at a concentration of 100 mmol/L. Then, ascorbic acid was added to the mixture at the same concentration and volume ratio of 1:1 under magnetic stirring. The solution changed from colorless to orange to dark orange, indicating the formation of SeNPs [33]. Moreover, the prepared SeNPs colloidal solution at the concentration of 50 mmol/L was used in this treatment process.

2.3. Treatment Method

The PP fabrics were treated using an infrared dyeing machine. The machine includes 12 beakers fixed in a rotating carrying wheel. Heating was obtained by IR, cooling by air, and automation by the microprocessor programmer DC4 F/R. The highest temperature used in this device was120 °C, the highest rate of heating was 2 °C per minute, and the highest rate of cooling was 6 °C per minute. First, a solution of SeNPs (50 mmol/L) was prepared. The PP fabrics were then immersed in the solution with liquor ratio (LR) of 1:50. Next, the device was set at the following treatment temperatures periods. The treatment was performed at different temperatures (70 °C, 100 °C and 120 °C) and different periods (1, 2 and 3 h). The treated PP fabrics were rinsed with distilled water and allowed to dry at room temperature after the treatment. The obtained fabrics were coded with SeNPs-PP fabrics.

2.4. Characterization

2.4.1. Characterization of Selenium Nanoparticles (SeNPs)

Transmission Electron Microscopy (TEM) Analysis

The size and morphology of SeNPs were characterized using JEM-2100 Transmission Electron Microscope with an acceleration voltage of 200 kV. A drop of colloidal solution containing SeNPs was dripped onto a carbon coated copper grid and dried at room temperature for TEM analysis.

X-ray Diffraction (XRD) Analysis

X-ray diffraction (XRD) analysis was conducted for synthesized SeNPs and SeNPs-PP fabrics using an X-ray diffractometer system (Bruker D8 ADVANCE, Karlsruhe, Germany). While SeNPs solution was dried at 130 °C until completely dryness before the XRD analysis.

UV/Vis Spectroscopy Analysis

SeNPs were further characterized via ultraviolet-visible spectrophotometer (Alpha-1860, Indianapolis, IN, USA), and their formation was confirmed by the maximum absorption peak which attributed to their surface plasmon resonance.

2.4.2. Characterization of Poly Propylene (PP) Fabrics

Scanning Electron Microscopy (SEM) Analysis

The surface morphology of blank PP and SeNPs-PP fabric was characterized via scanning electron microscope (JEOL JSM-6510LB with field emission gun, Tokyo, Japan). The deposition of SeNPs into PP fabrics was confirmed using surface energy dispersive x-ray (EDX) analysis unit (EDAX AMETEK analyzer) attached to SEM device.

Raman Spectroscopy Analysis

The types of bonds present in the blank PP and SeNPs-PP fabrics were determined using confocal Raman microscope (Jasco NRS-4500, Tokyo, Japan) which covered the range from 200 to 4000 cm^{-1}. Raman data acquisition and processing were performed using Jasco spectroscopy suite software.

Colorimetric Study

The colorimetric parameters such as lightness (L*), redness-greenness (a*), yellowness-blueness (b*) and the color uptake which is expressed as the color strength (K/S) of the obtained SeNPs-PP fabrics were determined using a spectrophotometer (CM3600A; Konica Minolta, Tokyo, Japan). K/S values were evaluated at the wavelength of maximum absorption (λ_{max}) of the color's reflectance curve at 390 nm.

Exhaustion of SeNPs into PP Fabric

The treatment colloidal solution was sampled before and after treatment to measure the exhaustion of SeNPs. The absorbance of the SeNPs colloidal solution was measured using an UV/Vis spectrophotometer (Alpha-1860, Indianapolis, IN, USA).

Physical Properties of SeNPs-PP Fabric

The fastness of the SeNPs-PP fabrics determines the fixation of SeNPs into the fabric. They were determined using AATCC (61-1972), (8-1972), and (16A-1972) [34] tests for washing, rubbing and lightfastness, respectively. The tensile strength tests of the PP and SeNPs-PP fabrics were performed using a universal testing machine (Tinius Olsen EN ISO

13934-1;1999-model H25KT) [35]. Additionally, the durability to washing was evaluated according to AATCC61(2A)-1996 test [36] after five washing cycles.

Cytotoxicity Test of SeNPs-PP Fabric

The cytotoxicity of SeNPs-PP fabric was tested against wi-38 cell line. This type of cells is diploid human cell line, including fibroblasts from lung tissue of a 3-month-gestation female fetus. The SeNPs-PP fabric treated at the optimum conditions was sterilized, cut, and plated on the bottom surface of a six-well tissue culture plate. The plate was inoculated with 1×10^5 cells/mL (100 µL/well) and incubated at 37 °C for 24 h. Additionally, the growth medium was decanted, and the cell monolayer was washed twice with washing media. Cells were checked for any physical signs of toxicity. Moreover, the tissue was picked up and 20 µL 3-(4,5-dimethylthiazol-2-yl)-2,5-diphenyltetrazolium bromide dye (MTT) prepared in phosphate buffer saline (BIO BASIC CANADA INC, Markham, Ontario, Canada) was added to each well at a concentration of 5mg/mL and shaken for 5 min. The wells were incubated at 37 °C and 5% CO_2 for 1–5 h. After dumping the media, the formazan as MTT metabolic product in the dry plate was resuspended in 200 µL dimethyl sulfoxide and shaken for 5 min. Then, the optical density was read at 560 nm, while the background was measured at 620 nm and subtracted.

Antibacterial Activity

The antibacterial activity of the SeNPs-PP fabric was evaluated using AATCC (147-2004) test [37]. Antibacterial tests were carried out against G+ve bacteria (*Staphylococcus aureus* and *Bacillus cereus*) as well as G-ve bacteria (*Escherichia coli* and *Pseudomonas aeruginosa*) and the growth inhibition zone (mm) was determined.

UV-Protection Properties

The UV protection factor (UPF) and UV-blocking activities of the SeNPs-PP fabric were determined using Standards Australia and Standards New Zealand (AS/NZS) 4399:1996 tests and UV protection properties were expressed as good, very good, or excellent for UPF values of 15–24, 25–39, and >40, respectively.

Electrical Conductivity Measurement

The electrical conductivity of both the PP and SeNPs-PP fabrics were measured using LRC-bridge (Hioki model 3531zHi Tester, Nagano, Japan).

3. Results and Discussion

3.1. Characterization of SeNPs

3.1.1. Transmission Electron Microscopy (TEM) Analysis

TEM micrographs confirmed the formation of spherical Se-NPs in the range of 31–79 nm. Furthermore, the synthesized SeNPs were well-dispersed with no aggregation and deformation as shown in Figure 1.

Histogram bins are 10-nm wide and centered at 35, 45, 55, 65 and 75 nm. All SeNPs with diameters from 40 nm to 50 nm were considered to have a size of 45 nm.

Furthermore, TEM was used to examine the adsorption of SeNPs as shown in Figure 1. The TEM images illustrated that SeNPs were sufficiently monodispersed and adsorbed on the SeNPs-PP fabric surface.

3.1.2. X-ray Diffraction (XRD) Analysis

XRD confirmed the formation of SeNPs and their deposition into the treated PP fabric based on the crystallinity of SeNPs. As illustrated in Figure 2, SeNPs in a colloidal solution or into the PP fabric surface were highly crystalline. Additionally, the diffraction peaks at 24.28°, 29.24°, 43.64° and 64.28° were corresponding to 100, 101, 102 and 210 crystal planes, respectively, based on the JCPDS 86-2246 international database [38]. Moreover, the peaks at 13,66, 16.56, 18.18, and 25.38 can be observed for PP fabric, corresponding to the planes of (110), (040), (130), and (060), respectively [39].

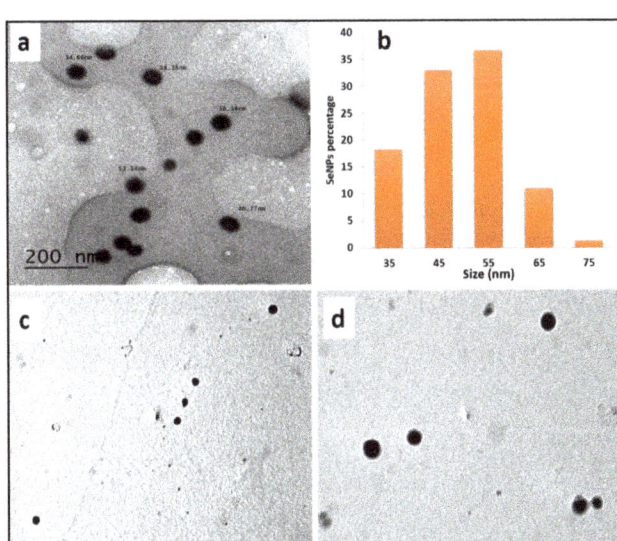

Figure 1. (a) TEM image (b) size distributions histogram of prepared SeNPs, (c,d) SeNPs-PP fabric surface.

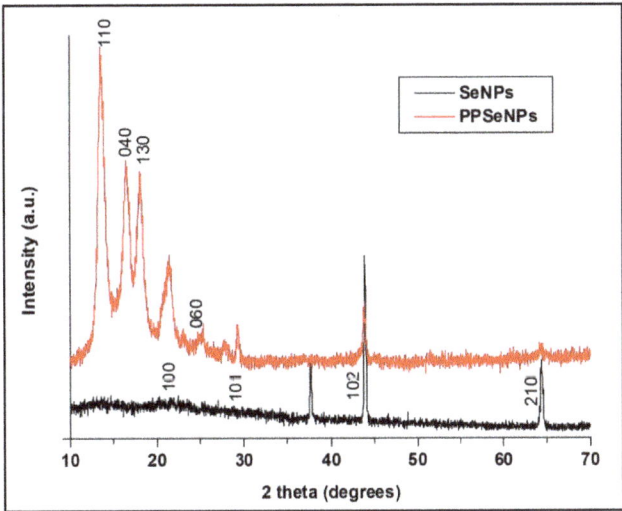

Figure 2. XRD patterns of the prepared SeNPs and SeNPs-PP fabric

3.1.3. UV/Vis Spectroscopy Analysis

The formation of SeNPs was confirmed from the UV/Vis spectra based on their SPRs. The solution changed from colorless to orange to dark orange, indicating the complete reduction of sodium hydrogen selenite to SeNPs [40].

As shown in Figure 3, the SeNPs colloidal solution showed an absorption peak at 263 nm, confirming the formation of the spherical SeNPs [41].

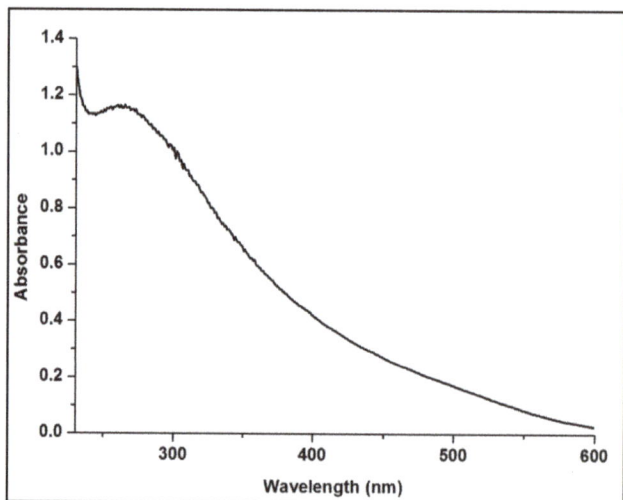

Figure 3. UV/Vis spectra of SeNPs at different concentrations.

3.2. Characterization of Poly Propylene (PP) Fabrics

3.2.1. Scanning Electron Microscopy (SEM) Analysis

The SEM micrographs of the PP fabric revealed that the surface was clear with clean scales and typical fibrous structure as displayed in Figure 4. On the other hand, the SEM micrographs of the SeNPs-PP fabric show a coated layer of SeNPs into the PP fabric. Additionally, SeNPs were well distributed into the fabric.

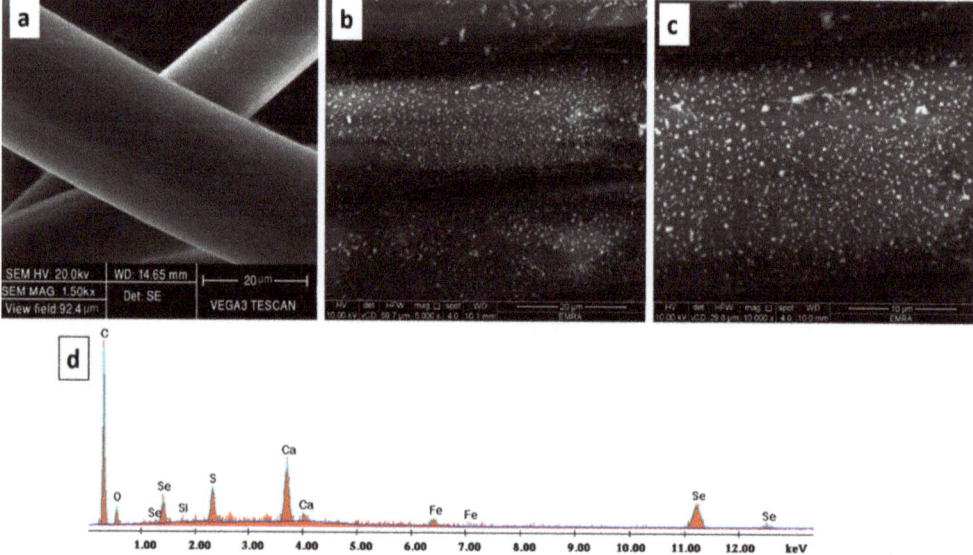

Figure 4. SEM micrographs of (**a**) the PP fabric, (**b**,**c**) SeNPs-PP fabric, (**d**) EDX analysis of the SeNPs-PP fabric.

The chemical elements found on the surface of the treated PP fabric were analyzed using EDX. The peaks around 1 and 11 Kev are attributed to SeNPs. The carbon and oxygen peaks were belonged to the native PP fabric. However, other elements were monitored at low concentration, such as Si, Ca, and Fe. Those traces of elements can be attributed to using IR-dyeing technique [42]. In addition, the Sulphur element was monitored at low concentration because EDX is an elemental detection technique with a certain small error. Both SEM micrographs and EDX analysis confirmed the deposition of SeNPs into the PP fabric surface as displayed in Figure 4.

3.2.2. Raman Spectroscopy

Raman analysis revealed the chemical bonds inside the PP fabric and between the SeNPs and PP surface. This analysis is important to compare the chemical structure of Se-NPs-PP fabric and PP fabric as shown in Figure 5. The peaks at 2960 as well as 2888 cm^{-1} were corresponding to the asymmetric stretch of methyl group [43]. While the peak at 984 cm^{-1} was associated with asymmetrical stretching of C-C bond [44]. Furthermore, Raman analysis revealed the presence of SeNPs into the SeNPs-PP fabric. The treated fabric showed an obvious peak at 236 cm^{-1}, corresponding to the symmetric stretching of SeNPs [45]. On the other hand, no peak was observed in this region in the case of PP fabric without SeNPs.

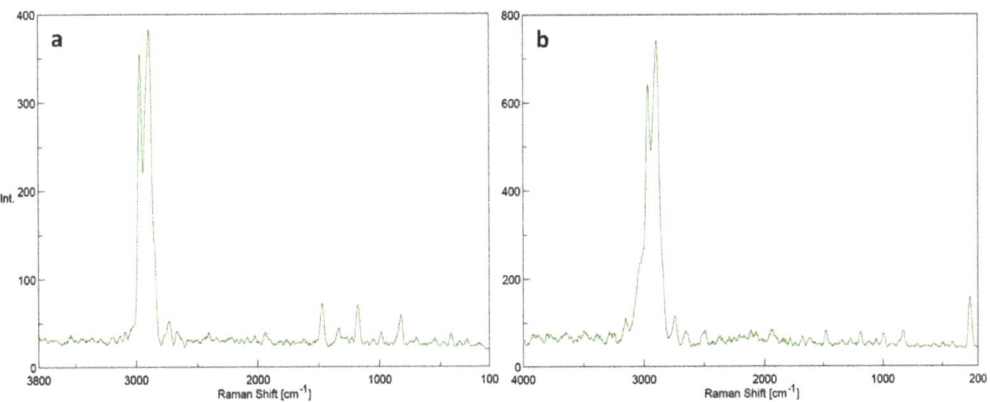

Figure 5. Raman spectrums of (**a**) PP fabric and (**b**) SeNPs-PP fabric.

3.2.3. Colorimetric Study

The color parameters of the SeNPs-PP fabric were analyzed using a Konica Minolta spectrophotometer (CM-3600A). Figure 6 and Table 1 show the L*a*b* values of the fabric, where (L*) values represent color lightness, (a*) is the red/green coordinate, and (b*) is the yellow/blue coordinate [46]. These values indicated that SeNPs-PP fabric is darker according to the color lightness values L*, less red and less yellow according to a*, b* values, respectively.

Effect of Treatment Time on Color Strength (K/S)

The relationship between the K/S value of the SeNPs-PP fabric and treatment time (1, 2, and 3 h) is shown in Figure 7. Notably, the *K/S* value of PP fabric treated with SeNPs (concentration of 50 mmole/L, L.R 1:50 and a temperature of 100 °C) increased with an increase in the treatment time. The increase in K/S value reflected the positive effect that increasing the treatment time had on the uniformity of PP adsorption of the SeNPs, and on the uniformity of the penetration and diffusion of the SeNPs into the fabric; these effects in turn contributed to an increase in the SeNPs uptake by the fabric [47,48]; which is indicated by the highest K/S value observed at treatment time of 3 h.

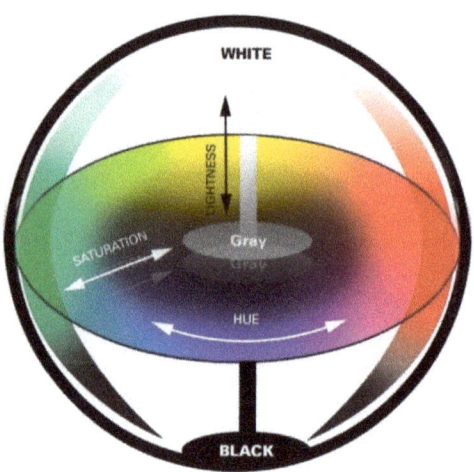

Figure 6. Lab colour space.

Table 1. Optical measurements.

Type	Sample	Colour Parameters					
		L*	a*	b*	C*	h	K/S
SeNPs-PP fabric		47.39	23.45	17.09	29.02	36.08	4.34

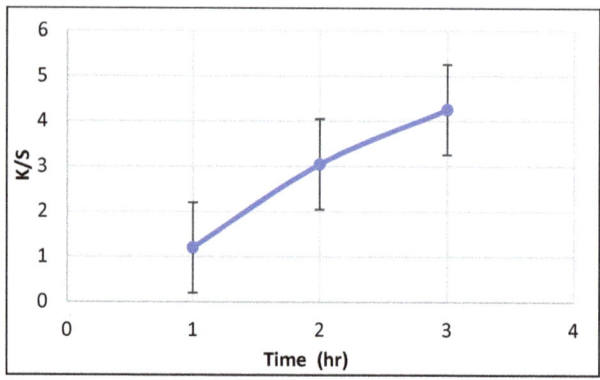

Figure 7. Effect of treatment time on color strength (K/S).

Effect of Treatment Temperature on Color Strength (K/S)

Figure 8 shows the relationship between treatment temperature and the color uptake (K/S) of the SeNPs-PP fabric with a (concentration of 50 mmol/L, L.R 1:50 and a treatment time of 3h). K/S increased linearly with an increase in temperature from 70°C to 120°C and increased considerably at low temperatures until the color approached to an equilibrium point above 100°C. The molecular structure opens, which facilitates the uptake of NPs as the temperature increases. Hence, a high K/S value is obtained. This can be attributed to an increase in temperature, which improves the macromolecular chains of PP. Moreover, large pores and/or channels suitable for NPs penetration and diffusion are formed. Hence, the optimum temperature was set at 100 °C [49–52].

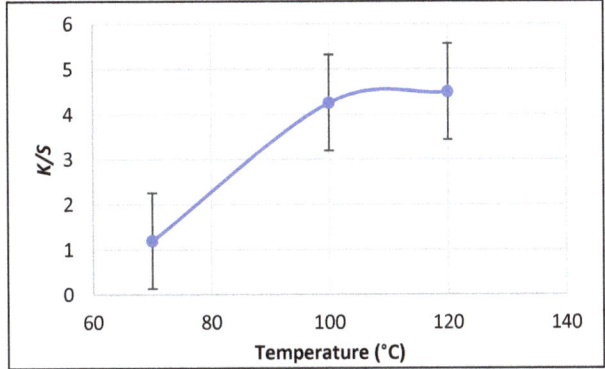

Figure 8. Effect of treatment temperature on color strength (K/S).

3.2.4. Exhaustion of SeNPs into PP Fabric

The treatment solution was sampled before and after treatment to measure the SeNPs exhaustion. Moreover, the absorbance of SeNPs solution was measured by using UV/VIS spectrophotometer-model: Alpha-1860. Figure 9 showed the absorbance of SeNPs concentration before and after exhaustion by PP fabric in the wavelength ranges from 200 to 700 nm. The absorption spectrum of SeNPs-PP fabric before exhaustion shows a sharp absorption band at 263 nm, indicating the presence of SeNPs. After exhaustion, remarkable decrease in the absorbance of the treatment solution can be attributed to the low ratio of SeNPs as it was absorbed by PP fabric.

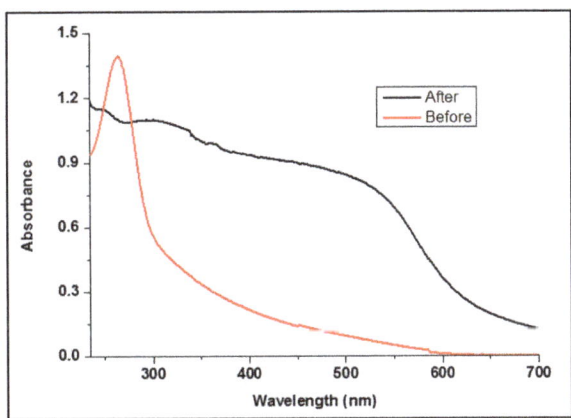

Figure 9. Absorbance of SeNPs before and after treatment.

3.2.5. Physical Properties of SeNPs-PP Fabric

The fastness of the SeNPs-PP fabric treated under optimum conditions was evaluated mainly in washing, rubbing, and lightfastness. In addition, the tensile strength (elongation and maximum force) was also evaluated. According to the results listed in Table 2, it can be concluded that there is a decrease in elongation and a little increase in maximum force of SeNPs-PP fabric without causing a significant damage to the structure of the yarn indicating no significant change between PP and SeNPs-PP fabrics. The washing and rubbing fastness were excellent, even after five washing cycles. Additionally, the light fastness of SeNPs-PP fabric was also found to be in the field of good to very good, this indicates that the fixation of SeNPs onto PP fabric may be attributed to the generation of metal chelates [53].

Table 2. Properties of the PP and SeNPs-PP fabrics under optimum conditions.

Sample	Wash Fastness		Rubbing Fastness		Light Fastness	Tensile Strength	
	St.	Alt.	Dry	Wet		Force, N	Elongation, %
PP fabric	-	-	-	-	-	1090	30.80
SeNPs-PP fabric	5	5	5	4–5	5	1102	27.36
SeNPs-PP fabric after 5 washing cycles (durability test)	5	5	5	4–5	5	-	-

3.2.6. Cytotoxicity of the SeNPs-PP Fabric

The cytotoxicity of the SeNPs-PP fabric was evaluated against healthy human cells (wi-38) by the MTT assay. The viability of cells and the cytotoxicity of SeNPs-PP fabric were evaluated against wi-38 cell lines. The viability of cells of SeNPs-PP fabric was 99.46% of that of negative control. Whereas over 70% is the mean relative cell viability [54]

3.2.7. Antibacterial Activity of the SeNPs-PP Fabric

Table 3 lists the antibacterial activity of the SeNPs-PP fabric using four bacterial strains, (*Staphylococcus aureus* and *Bacillus cereus*) as Gram-positive bacteria and (*Escherichia coli* and *Pseudomonas aeruginosa*) as Gram-negative bacteria. Tetracycline and ciprofloxacin were used as standard drugs. The results reveal that the SeNPs-PP fabric exhibits excellent antibacterial activity against *Escherichia coli, Bacillus cereus*, as well as *Staphylococcus aureus* and very good antibacterial activity against *Pseudomonas aeruginosa*, which is indicated by a clear zone diameter of bacterial colonies. The obtained results indicate the presence of broad-spectrum antibacterial activity. The mechanism of action of SeNPs on bacteria is still unclear. In this study, we suggest the following phenomena: (a) the release of ions along with physical interaction with the bacterial cell wall peptidoglycan layer damages the double-stranded structure of DNA. (b) the formation of reactive oxygen species and inhibition of DNA replication. (c)nanoparticles can be in better contact with bacterial or fungal cells than colloidal form [53].

Table 3. Clear zone (mm) of the PP and SeNPs-PP fabrics.

Substrate	Antibacterial Activity Diameter of Clear Zone (mm)			
	Staphylococcus aureus (G+)	*Bacillus cereus* (G+)	*Escherichia coli* (G-)	*Pseudomonas aeruginosa* (G-)
Ciprofloxacin	24	15	23	17
Tetracycline	21	14	20	15
Blank	–	–	–	–
SeNPs-PP fabric	20.9	22.7	23.2	11.3

3.2.8. UV-Protection Properties of the SeNPs-PP Fabric

Table 4 lists the results of the UV light protection characterization. The SeNPs-PP fabric effectively blocked UV radiation; based on the AATCC test criteria: if UPF of any fabric is more than 40as the fabric is a UV-defensive material [55].

Table 4. UPF of the PP and SeNPs-PP fabrics.

Sample	UVA 315–400 nm	UVB 290–315 nm	UPF Value
Blank PP fabric	35.89	29.49	3.20
SeNPs-PP fabric	0.11	0.11	920.19

3.2.9. Electrical Conductivity Measurement

The treatment of PP fabrics with SeNPs led to a slight increase in the electrical conductivity of PP fabrics. Samples treated with the optimum concentration of SeNPs showed EC value of $5.84 \times 10^{-11} \Omega^{-1} \text{Cm}^{-1}$ compared to untreated fabric which had EC value of $1.06 \times 10^{-11} \Omega^{-1} \text{cm}^{-1}$.

4. Conclusions

In this paper, we propose a novel approach for coloring and incorporating new functionalities to PP fabrics via one-step process using SeNPs. PP fabrics were colored from light to dark orange depending on the treatment time and concentration of sodium hydrogen selenite. The obtained results show that the deposition of SeNPs into PP fabric is accompanied by a considerable improvement in UV-protection. The obtained colored fabrics effectively blocked UV radiations, providing excellent UV-protection. Additionally, the treated fabrics exhibited outstanding washing, rubbing and light fastness. Moreover, the colored PP fabric showed excellent antibacterial activity against *Staphylococcus aureus*, *Bacillus cereus*, and *Escherichia coli* and very good antibacterial activity against *Pseudomonas aeruginosa* compared with the standard drugs such as tetracycline and ciprofloxacin. The tensile strength of the colored fabrics increased slightly accompanied by a slight decrease in elongation. This novel, and economical approach can be employed in the industry for coloration and multifunctionalization of PP fabrics instead of traditional dyeing and finishing processes.

Author Contributions: Conceptualization, T.A. and S.A.A.; methodology, S.A.A.; software, S.A.A. validation, S.M.R., K.S.-A. and T.A.; formal analysis, K.S.-A.; investigation, S.M.R.; resources, S.M.R.; data curation, S.A.A.; writing—original draft preparation, S.A.A.; writing—review and editing, T.A. and M.R.P.; visualization, M.R.P.; supervision, T.A.; project administration, T.A. All authors have read and agreed to the published version of the manuscript.

Funding: This research received no external funding.

Institutional Review Board Statement: Not applicable.

Informed Consent Statement: Not applicable.

Data Availability Statement: The data presented in this study are available on request from the corresponding author.

Conflicts of Interest: The authors declare no conflict of interest.

References

1. Marković, D.; Tseng, H.-H.; Nunney, T.; Radoičić, M.; Ilic-Tomic, T.; Radetić, M. Novel antimicrobial nanocomposite based on polypropylene non-woven fabric, biopolymer alginate and copper oxides nanoparticles. *Appl. Surf. Sci.* **2020**, *527*, 146829. [CrossRef]
2. Gupta, D.; Khare, S.K.; Laha, A. Antimicrobial properties of natural dyes against Gram-negative bacteria. *Color. Technol.* **2004**, *120*, 167–171. [CrossRef]
3. Wang, Q.; Barnes, L.-M.; Maslakov, K.I.; Howell, C.A.; Illsley, M.J.; Dyer, P.; Savina, I.N. In situ synthesis of silver or selenium nanoparticles on cationized cellulose fabrics for antimicrobial application. *Mater. Sci. Eng. C* **2021**, *121*, 111859. [CrossRef]
4. Beyth, N.; Houri-Haddad, Y.; Domb, A.; Khan, W.; Hazan, R. Alternative Antimicrobial Approach: Nano-Antimicrobial Materials. *Evid. Based Complement. Altern. Med.* **2015**, *2015*, 246012. [CrossRef] [PubMed]
5. Abou Elmaaty, T.; Abdeldayem, S.; Elshafai, N. Simultaneous Thermochromic Pigment Printing and Se-NP Multifunctional Finishing of Cotton Fabrics for Smart Childrenswear. *Cloth. Text. Res. J.* **2020**, *38*, 182–195. [CrossRef]

6. Abou Elmaaty, T.; Mostafa, S.; Nasr Eldin, S.; Elgamal, G.M. One Step Thermochromic Pigment Printing and Ag NPs Antibacterial Functional Finishing of Cotton and Cotton/PET Fabrics. *Fibers Polym.* **2018**, *19*, 2317–2323. [CrossRef]
7. Abou Elmaaty, T.; Abdelaziz, E.; Nasser, D.; Abdelfattah, K.; Elkadi, S.; El-Nagar, K.I. Microwave and nanotechnology advanced solutions to improve ecofriendly cotton's coloration and performance properties. *Egypt. J. Chem.* **2018**, *61*, 493–502. [CrossRef]
8. Abou Elmaaty, T.; Elnnagar, K.; Raouf, S.; Abdelfattah, K.; El-Kadi, S.; Abdelaziz, E. One-step green approach for functional printing and finishing of textiles using silver and gold NPs. *RSC Adv.* **2018**, *8*, 25546–25557. [CrossRef]
9. Abou Elmaaty, T.; Mandour, B. ZnO and TiO_2 Nanoparticles as Textile Protecting Agents against UV Radiation: A Review. *Asian J. Chem. Sci.* **2018**, *4*, 1–14. [CrossRef]
10. Gold, K.; Slay, B.; Knackstedt, M.; Gaharwar, A.K. Antimicrobial Activity of Metal and Metal-Oxide Based Nanoparticles. *Adv. Ther.* **2018**, *1*, 1700033. [CrossRef]
11. Roy, A.; Bulut, O.; Some, S.; Mandal, A.K.; Yilmaz, M.D. Green synthesis of silver nanoparticles: Biomolecule-nanoparticle organizations targeting antimicrobial activity. *RSC Adv.* **2019**, *9*, 2673–2702. [CrossRef]
12. Dastjerdi, R.; Montazer, M. A review on the application of inorganic nano-structured materials in the modification of textiles: Focus on anti-microbial properties. *Colloids Surf. B* **2010**, *79*, 5–18. [CrossRef]
13. Simončič, B.; Klemenčič, D. Preparation and performance of silver as an antimicrobial agent for textiles: A review. *Text. Res. J.* **2015**, *86*, 210–223. [CrossRef]
14. Radetić, M. Functionalization of textile materials with silver nanoparticles. *J. Mater. Sci.* **2013**, *48*, 95–107. [CrossRef]
15. Radetić, M.; Marković, D. Nano-finishing of cellulose textile materials with copper and copper oxide nanoparticles. *Cellulose* **2019**, *26*, 8971–8991. [CrossRef]
16. Yip, J.; Liu, L.; Wong, K.-H.; Leung, P.H.M.; Yuen, C.-W.M.; Cheung, M.-C. Investigation of antifungal and antibacterial effects of fabric padded with highly stable selenium nanoparticles. *J. Appl. Polym. Sci.* **2014**, *131*. [CrossRef]
17. Biswas, D.P.; O'Brien-Simpson, N.M.; Reynolds, E.C.; O'Connor, A.J.; Tran, P.A. Comparative study of novel in situ decorated porous chitosan-selenium scaffolds and porous chitosan-silver scaffolds towards antimicrobial wound dressing application. *J. Colloid Interface Sci.* **2018**, *515*, 78–91. [CrossRef] [PubMed]
18. Yu, B.; Zhang, Y.; Zheng, W.; Fan, C.; Chen, T. Positive Surface Charge Enhances Selective Cellular Uptake and Anticancer Efficacy of Selenium Nanoparticles. *Inorg. Chem.* **2012**, *51*, 8956–8963. [CrossRef]
19. Forootanfar, H.; Adeli-Sardou, M.; Nikkhoo, M.; Mehrabani, M.; Amir-Heidari, B.; Shahverdi, A.R.; Shakibaie, M. Antioxidant and cytotoxic effect of biologically synthesized selenium nanoparticles in comparison to selenium dioxide. *J. Trace Elem. Med. Biol.* **2014**, *28*, 75–79. [CrossRef]
20. Shakibaie, M.; Forootanfar, H.; Golkari, Y.; Mohammadi-Khorsand, T.; Shakibaie, M.R. Anti-biofilm activity of biogenic selenium nanoparticles and selenium dioxide against clinical isolates of *Staphylococcus aureus*, *Pseudomonas aeruginosa*, and *Proteus mirabilis*. *J. Trace Elem. Med. Biol.* **2015**, *29*, 235–241. [CrossRef] [PubMed]
21. Hariharan, H.; Al-Harbi, N.; Karuppiah, P.; Rajaram, S.K. Microbial synthesis of selinium nanocomposite using *Saccharomyces cerevisiae* and its antimicrobial activity against pathogens causing nosocomial infection. *Chalcog. Lett.* **2012**, *9*, 509–515.
22. Shakibaie, M.; Salari Mohazab, N.; Ayatollahi Mousavi, S.A. Antifungal Activity of Selenium Nanoparticles Synthesized by Bacillus species Msh-1 against *Aspergillus fumigatus* and *Candida albicans*. *Jundishapur J. Microbiol.* **2015**, *8*, e26381. [CrossRef]
23. Beheshti, N.; Soflaei, S.; Shakibaie, M.; Yazdi, M.H.; Ghaffarifar, F.; Dalimi, A.; Shahverdi, A.R. Efficacy of biogenic selenium nanoparticles against *Leishmania major*: In vitro and in vivo studies. *J. Trace Elem. Med. Biol.* **2013**, *27*, 203–207. [CrossRef] [PubMed]
24. Gawish, S.M.; Mosleh, S.E.-S.; Ramadan, A. Review Improvement of Polypropylene Properties by Irradiation/Grafting and Other Modifications. *Egypt. J. Chem.* **2019**, *62*, 29–48. [CrossRef]
25. Gawish, S.; Ramadan, A.; Mosleh, S. Improvement of Polypropylene (PP) Dyeing by Modification Methods. *Egypt. J. Chem.* **2018**, *62*, 49–62. [CrossRef]
26. Tao, G.; Gong, J.; Lu, J.; Sue, H.-J.; Bergbreiter, D.E. Surface Functionalized Polypropylene: Synthesis, Characterization, and Adhesion Properties. *Macromolecules* **2001**, *34*, 7672–7679. [CrossRef]
27. Goddard, J.M.; Hotchkiss, J.H. Polymer surface modification for the attachment of bioactive compounds. *Prog. Polym. Sci.* **2007**, *32*, 698–725. [CrossRef]
28. Bandopadhay, D.; Tarafdar, A.; Panda, A.B.; Pramanik, P. Surface modification of low-density polyethylene films by a novel solution base chemical process. *J. Appl. Polym. Sci.* **2004**, *92*, 3046–3051. [CrossRef]
29. Perkas, N.; Shuster, M.; Amirian, G.; Koltypin, Y.; Gedanken, A. Sonochemical immobilization of silver nanoparticles on porous polypropylene. *J. Polym. Sci. Part A Polym. Chem.* **2008**, *46*, 1719–1729. [CrossRef]
30. Gawish, S.M.; Mosleh, S. Antimicrobial Polypropylene Loaded by Silver Nano Particles. *Fibers Polym.* **2020**, *21*, 19–23. [CrossRef]
31. Elmaaty, T.A.; Raouf, S.; Sayed-Ahmed, K. Novel One Step Printing and Functional Finishing of Wool Fabric Using Selenium Nanoparticles. *Fibers Polym.* **2020**, *21*, 1983–1991. [CrossRef]
32. Brandrup, J.; Immergut, E.H.; Grulke, E.A.; Abe, A.; Bloch, D.R. *Polymer Handbook*; Wiley: New York, NY, USA, 1999; p. 89.
33. Malhotra, S.; Jha, N.; Desai, K. A superficial synthesis of selenium nanospheres using wet chemical approach. *Int. J. Nanotechnol. Appl.* **2013**, *3*, 7–14.
34. American Association of Textile Chemists and Colorists. *AATTCC 16A-1972: Colorfastness to Light: Carbon-arc Lamp*; American Association of Textile Chemists and Colorists: Research Triangle Park, NC, USA, 1972.

35. International Organization for Standardization. *ISO 13934-1:1999: Tensile Properties of Fabrics*; ISO: Vernier, Switzerland, 1999.
36. American Association of Textile Chemists and Colorists. *AATTCC 61(2A)-1996: Colorfastness to Laundering*; American Association of Textile Chemists and Colorists: Research Triangle Park, NC, USA, 1996.
37. American Association of Textile Chemists and Colorists. *AATTCC 147-2004: Antimicrobial Activity Assessment of Textile Materials*; American Association of Textile Chemists and Colorists: Research Triangle Park, NC, USA, 2004.
38. Vieira, A.; Stein, E.; Andreguetti, D.; Cebrián-Torrejón, G.; Domènech, A.; Colepicolo, P.; Da Costa Ferreira, A.M. "Sweet Chemistry": A Green Way for Obtaining Selenium Nanoparticles Active against Cancer Cells. *J. Braz. Chem. Soc.* **2017**, *28*, 2021–2027. [CrossRef]
39. Wang, S.; Ajji, A.; Guo, S.; Xiong, C. Preparation of Microporous Polypropylene/Titanium Dioxide Composite Membranes with Enhanced Electrolyte Uptake Capability via Melt Extruding and Stretching. *Polymers* **2017**, *9*, 110. [CrossRef]
40. Alagesan, V.; Venugopal, S. Green Synthesis of Selenium Nanoparticle Using Leaves Extract of Withania somnifera and Its Biological Applications and Photocatalytic Activities. *BioNanoScience* **2019**, *9*, 105–116. [CrossRef]
41. Satgurunathan, T.; Bhavan, P.; Komathi, S. Green synthesis of selenium nanoparticles from sodium selenite using garlic extract and its enrichment on Artemia nauplii to feed the fresh water prawn Macrobrachium rosenbergii post-larvae. *Res. J. Chem. Environ.* **2017**, *21*, 1–12.
42. El-Kheir, A.; El-Ghany, N.; Fahmy, M.; Aboras, S.; El-Gabry, L. Functional Finishing of Polyester Fabric Using Bentonite Nano-Particles. *Egypt. J. Chem.* **2019**, *63*. [CrossRef]
43. Arruebarrena de Báez, M.; Hendra, P.J.; Judkins, M. The Raman spectra of oriented isotactic polypropylene. *Spectrochim. Acta Part A Mol. Spectrosc.* **1995**, *51*, 2117. [CrossRef]
44. Martin, J.; Ponçot, M.; Hiver, J.M.; Bourson, P.; Dahoun, A. Real-time Raman spectroscopy measurements to study the uniaxial tension of isotactic polypropylene: A global overview of microstructural deformation mechanisms. *J. Raman Spectrosc.* **2013**, *44*, 776–784. [CrossRef]
45. Bashiri Rezaie, A.; Montazer, M.; Mahmoudi Rad, M. Low toxic antibacterial application with hydrophobic properties on polyester through facile and clean fabrication of nano copper with fatty acid. *Mater. Sci. Eng. C* **2019**, *97*, 177–187. [CrossRef]
46. Khalifa, M.E.; Abdel-Latif, E.; Gobouri, A.A. Disperse Dyes Based on 5-Arylazo-thiazol-2-ylcarbamoyl-thiophenes: Synthesis, Antimicrobial Activity and Their Application on Polyester. *J. Heterocycl. Chem.* **2015**, *52*, 674–680. [CrossRef]
47. Long, J.-J.; Ma, Y.-Q.; Zhao, J.-P. Investigations on the level dyeing of fabrics in supercritical carbon dioxide. *J. Supercrit. Fluids* **2011**, *57*, 80–86. [CrossRef]
48. Hou, A.; Dai, J. Kinetics of dyeing of polyester with CI Disperse Blue 79 in supercritical carbon dioxide. *Coloration Technol.* **2005**, *121*, 18–20. [CrossRef]
49. Al-Etaibi, A.M.; Alnassar, H.S.; El-Apasery, M.A. Dyeing of polyester with disperse dyes: Part 2. Synthesis and dyeing characteristics of some azo disperse dyes for polyester fabrics. *Molecules* **2016**, *21*, 855. [CrossRef]
50. Farizadeh, K.; Yazdanshenas, M.; Montazer, M.; Malek, R.; Rashidi, A. Kinetic studies of adsorption of madder on wool using various models. *Text. Res. J.* **2010**, *80*, 847–855. [CrossRef]
51. Kim, S.; Wairkar, Y.P.; Daniels, R.W.; DiAntonio, A. The novel endosomal membrane protein Ema interacts with the class C Vps–HOPS complex to promote endosomal maturation. *J. Cell Biol.* **2010**, *188*, 717–734. [CrossRef]
52. Miah, L.; Ferdous, N.; Azad, D. Textiles Material Dyeing with Supercritical Carbon Dioxide (CO_2) without using Water. *Chem. Mater. Res.* **2013**, *3*, 38–40.
53. Pandit, P.; Teli, M.D.; Therani Nadathur, G.; Maiti, S.; Singha, K.; Maity, S. Green synthesis of nanoparticle and its application on cotton fabric using Sterculia foetida fruit shell extract. *J. Text. Eng. Fash. Technol.* **2020**, *6*. [CrossRef]
54. Kangwansupamonkon, W.; Lauruengtana, V.; Surassmo, S.; Ruktanonchai, U. Antibacterial effect of apatite-coated titanium dioxide for textiles applications. *Nanomed. Nanotechnol. Biol. Med.* **2009**, *5*, 240–249. [CrossRef]
55. Mousa, M.A.; Khairy, M. Synthesis of nano-zinc oxide with different morphologies and its application on fabrics for UV protection and microbe-resistant defense clothing. *Text. Res. J.* **2020**, *90*, 2492–2503. [CrossRef]

Article

Electromagnetic Wave-Absorbing and Bending Properties of Three-Dimensional Honeycomb Woven Composites

Li-Hua Lyu [1,2,*], Wen-Di Liu [2] and Bao-Zhong Sun [1]

1. Key Laboratory of High Performance Fibers & Products, Ministry of Education, Donghua University, Shanghai 201620, China; sunbz@dhu.edu.cn
2. School of Textile and Material Engineering, Dalian Polytechnic University, Dalian 116034, China; 18108210003147@xy.dlpu.edu.cn
* Correspondence: lvlh@dlpu.edu.cn

Abstract: To avoid the delamination of the traditional three-dimensional (3-D) honeycomb electromagnetic (EM) absorbing composites and improving the defects of low mechanical properties, the 3-D honeycomb woven fabrics were woven on the ordinary loom by practical design. The fabrication of 3-D honeycomb woven EM absorbing composites was based on carbon black/carbonyl iron powder/basalt fiber/carbon fiber/epoxy resin (CB/CIP/BF/CF/EP) by the vacuum-assisted resin transfer molding (VARTM) process. A CB/CIP composite absorbent study showed that CB/CIP composite absorbent belongs to a magnetic loss type absorbent. Adding CB/CIP significantly improved the absorption performance of composite, increased the absorption peak and the effective absorption bandwidth (EAB), but the bending performance decreased. The normalization analysis results showed that when the thickness was 15 mm, the mechanical properties and EM wave-absorbing properties of the 3-D honeycomb woven composite were the best matches. The morphological characteristics and displacement load curves of the composite after fracture were analyzed. The bending failure modes were brittle fracture of the fiber bundle, matrix cracking, and typical shear failure. Despite the above failure mechanism, the 3-D honeycomb woven EM absorbing composites still has good integrity without delamination.

Keywords: 3-D honeycomb woven fabric; CB/CIP; mechanical property; EM wave-absorbing property

Citation: Lyu, L.-H.; Liu, W.-D.; Sun, B.-Z. Electromagnetic Wave-Absorbing and Bending Properties of Three-Dimensional Honeycomb Woven Composites. *Polymers* **2021**, *13*, 1485. https://doi.org/10.3390/polym13091485

Academic Editor: Tarek M. Abou Elmaaty

Received: 5 April 2021
Accepted: 30 April 2021
Published: 5 May 2021

Publisher's Note: MDPI stays neutral with regard to jurisdictional claims in published maps and institutional affiliations.

Copyright: © 2021 by the authors. Licensee MDPI, Basel, Switzerland. This article is an open access article distributed under the terms and conditions of the Creative Commons Attribution (CC BY) license (https://creativecommons.org/licenses/by/4.0/).

1. Introduction

The problem of electromagnetic (EM) interference and radiation has stimulated the research interest in EM absorbing composites with thin thickness, low density, broadband and strong absorption [1]. The research of EM absorbing composites has gradually turned to structural EM absorbing composites, which have the functions of load-bearing and a broadband EM energy absorbing capability. At present, the research of structural absorbing composites, such as foam EM absorbing composites and honeycomb EM absorbing composites, is being studied to overcome the disadvantages of traditional EM absorbing composites, such as high density, weak absorption and narrow bandwidth [2]. Compared with other structures, the honeycomb structure can not only be used as a part of the main load-bearing structure, but also as a carrier of EM absorbing composites and absorbing medium [3,4]. Li et al. [5] prepared carbon-coated honeycomb absorbing composites and analyzed in detail the effects of honeycomb aperture, honeycomb height, coating thickness, dielectric constant, and dielectric loss factor on EM absorption. Wang et al. [6] integrated shorted carbon fibers into honeycomb frames as a dual-functional material for radar absorption and structural reinforcement, which showed excellent EM absorption at frequencies ranging from 2–18 GHz. Sun et al. [7] used a honeycomb structure of aramid paper impregnated with EM absorbent as the core layer and quartz fiberboard as the skin to form a composite sandwich structure with EM absorption ability bonding. The experimental results showed that the minimum reflection loss (RLmin) was −29.5 dB, and the EAB was 13.1 GHz.

At present, honeycomb structure EM absorbing composites are mostly prepared by the lamination method, which has some disadvantages which were under high temperature and high humidity environments or alternating external forces, the honeycomb structure EM absorbing composites are easy to break [8–10]. Therefore, the traditional honeycomb structure EM absorbing composites have difficulty achieving load-bearing and a broadband EM energy absorbing capability.

However, three-dimensional (3-D) honeycomb woven composites are composites which are made of 3-D honeycomb woven fabric as reinforcement and resin or other polymers as the matrix. Three-dimensional honeycomb woven fabric is a kind of 3-D structure textile material. In the fabric structure of the textile material, the concept of vertical yarn was introduced along the thickness direction of the material, and the vertical yarn was used to connect the warp yarn and the weft yarn. There were fiber bundles and reinforcement structures through the thickness direction to enhance the integrity of the textile materials. Therefore, 3-D honeycomb woven composites have good integrity and structural stability and can solve the interlayer problems of traditional honeycomb structure EM absorbing composites [11]. Three-dimensional honeycomb woven fabrics have been applied in a number of practical applications, for example fibrous porous media [12]. Bayraktar et al. [13] prepared a 3-D honeycomb woven composite and carried out low-speed impact tests on fabric samples using a drop-weight impact tester. The experimental results showed that the energy absorbed by the honeycomb structure was more significant than that absorbed by the plate sample. The honeycomb woven fabric showed good mechanical properties. Lv et al. [14] prepared 3-D honeycomb structure composites using glass fiber filament yarn/BF filament yarn on the ordinary loom and tested their bending properties, providing a reference structural optimization design and performance analysis of 3-D honeycomb woven composites. Zahid et al. [15] wove three kinds of 3-D honeycomb woven fabrics with different honeycomb sizes, and the breaking strength and elongation at break of the 3-D honeycomb woven fabrics were measured. The effect of honeycomb size on the mechanical properties should be considered in the design and weaving of 3-D honeycomb woven fabrics.

Current studies have shown that 3-D honeycomb woven composites have good anti-delamination abilities. However, there was no research on the EM absorbing property of 3-D honeycomb woven composites. Therefore, the low-cost design and preparation of 3-D honeycomb woven EM absorbing composites with the integration of load-bearing and absorbing structures and functions can realize the collaborative design of mechanical properties and EM absorbing properties the composite.

Therefore, in this study, three different thicknesses (7.5 mm, 15 mm, 22.5 mm) of 3-D honeycomb fabrics with ordinary looms were woven through practical design. The CB/CIP/EP as the matrix and the 3-D honeycomb woven EM absorbing composites were prepared by the VARTM process. Then, the coaxial method was used to study the EM properties of CB/CIP. The 3-D honeycomb woven EM absorbing composite used the United States Naval Research Laboratory (NRL) arch method to test the RL on 2–18 GHz. Next, the bending properties were tested by a microcomputer-controlled electronic universal testing machine. Finally, it discussed particles and absorbent material thickness absorbing the composites' performance and the influence of mechanical properties, and experimental results obtained from the failure modes. An effective method to coordinate the design of absorbing and mechanical properties was proposed.

2. Materials and Methods

2.1. Materials and Equipment

EP JC-02A and the solidification reagent JC-02B (Changshu Jiafa Chemical Co., Ltd., Changshu City, China). 800 tex CF filament SYT49 (Zhongfu Shenying Carbon Fiber Co., Ltd., Lianyungang City, China). 800 tex BF filament (Zhejiang Shijin Basalt Fiber Co., Ltd.,

Hangzhou City, China). Ordinary loom SGA 598 (Jiangyin Tong Yuan spinning machine Co., Ltd., Wuxi City, China). The universal testing machine TH-8102S (Suzhou Tuobo Machinery Equipment Co., Ltd., Suzhou City, China).

2.2. Preform Design and Weaving

The warp structural drawings of 3-D honeycomb woven fabrics with three different layers are shown in Figure 1. As shown in Figure 1, the bottom warp and weft yarns were made entirely of 800 tex CF filament, and the rest of the warp and weft yarns were wholly made of 800 tex BF filament.

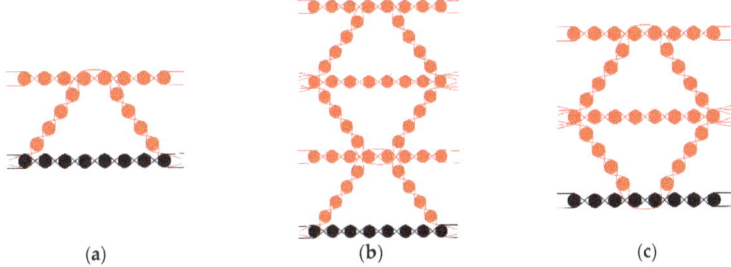

Figure 1. (**a**) Warp structural drawings of 7.5 mm 3D honeycomb woven fabric; (**b**) warp structural drawings of 15 mm 3D honeycomb woven fabric; (**c**) warp structural drawings of 22.5 mm 3D honeycomb woven fabric.

The chain drafts of 3-D honeycomb woven fabrics with three different thicknesses were drawn up according to the warp structural drawings, and they are shown in Figure 2. The weaving parameters of 3-D honeycomb woven fabrics are shown in Table 1.

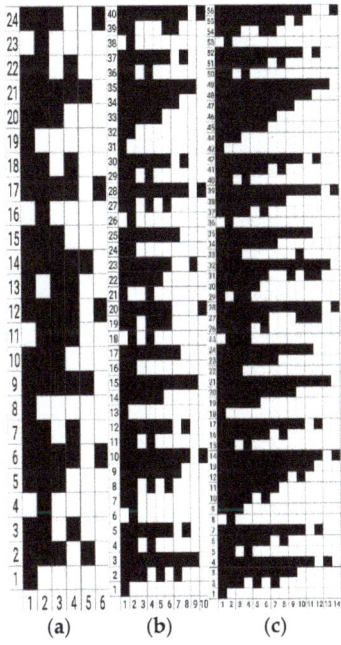

Figure 2. (**a**) Chain draft of 7.5 mm 3-D honeycomb woven fabric; (**b**) chain draft of 15 mm 3-D honeycomb woven fabric; (**c**) chain draft of 22.5 mm 3-D honeycomb woven fabric.

Table 1. Weaving parameters of 3-D honeycomb woven fabrics.

Thickness (mm)	Linear Density/Tex Warp/Weft Yarns	Layer Number of Yarns	Weaving Density (Yarn/10 cm)	
			Warp Density	Weft Density
7.5		1	180	
15	800	2	300	95
22.5		3	420	

2.3. Fabrication of the 3-D Honeycomb Woven EM Absorbing Composites

The ratio of the EP, CIP, solidification reagent, and CB was 4:4:3.2:0.03. VARTM manufactured the 3-D honeycomb woven EM absorbing composites. Figure 3a showed the schematic diagram of the VARTM process. The CB, CIP, and EP were mixed in a beaker and mechanically stirred until the CB/CIP could be uniformly dispersed in the EP without significant settling. After vacuum defoaming for 1 h in a vacuum drying oven at 60 °C, the CB/CIP/EP mixed solution without bubbles was obtained. The process involved placing the one-layer of 3-D honeycomb woven fabric (200 mm × 200 mm × 7.5 mm), the two-layer of 3-D honeycomb woven fabric (200 mm × 200 mm × 15 mm), and the three-layer of 3-D honeycomb woven fabric (200 mm × 200 mm × 22.5 mm) into their molds, closing the molds, and checking for leaks. Then, the resin solution was injected into the mold. The injection process was continued until a sufficient resin volume was seen in the resin trap to indicate that the mold had been completely filled with resin. The mold was isolated from the resin pot and the resin trap and then put into an air-circulating oven. The manufacturer-recommended cure cycle was employed: the first step of the cure cycle was 2 h at 90 °C, with the second step being 1 h at 130 °C, and the final step being 4 h at 150 °C, as shown in Figure 3b. The 3-D honeycomb woven EM absorbing composite without adding CB/CIP, with a thickness of 7.5 mm, was recorded as S0. The 3-D honeycomb woven EM absorbing composite with added CB/CIP, with a thickness of 7.5 mm, was recorded as S1. The 3-D honeycomb woven EM absorbing composite with added CB/CIP, with a thickness of 15 mm, was recorded as S2. The 3-D honeycomb woven EM absorbing composite with added CB/CIP, with a thickness of 22.5 mm, was recorded as S3.

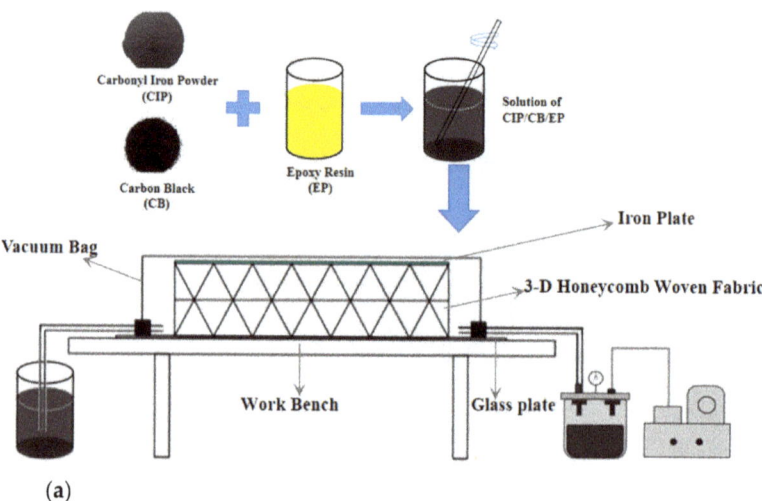

(a)

Figure 3. Cont.

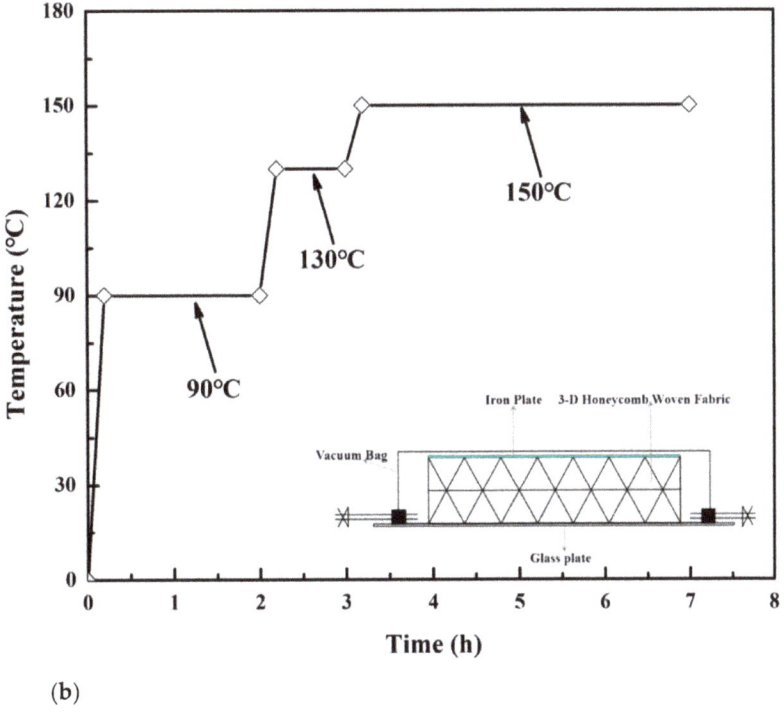

(b)

Figure 3. (a) Schematic diagram of VARTM process; (b) high-temperature curing process diagram of the 3-D honeycomb woven EM absorbing composite.

2.4. Characterization Methods

The EM parameters of the CB/CIP were tested by the coaxial method. The specimens were prepared by homogeneously mixing the CB/CIP with the paraffin in a mass ratio of 0.03:4:4; the mixture was pressed into a toroidal shape with an inner diameter of 3.04 mm and an outer diameter of 7 mm. The schematic of the experimental setup is shown in Figure 4a.

The morphology of the absorbent was observed with the scanning electron microscope JSM-7800F (JEOL Ltd., Guangzhou City, China) in the high-vacuum. The applied accelerating voltage was 15 kV and the working distance was 10 mm.

The United States NRL arch measurement system was used to evaluate the broadband EM characteristics of the 3-D honeycomb woven EM absorbing composites with different layers with a reference standard of GJB2038-2011; the schematic of the experimental setup is shown in Figure 4b.

Three-point bending with a universal testing machine followed GB/T 9341-2008. The loading rates were 10 mm/min for the bending test. The schematic of the experimental setup is shown in Figure 4c.

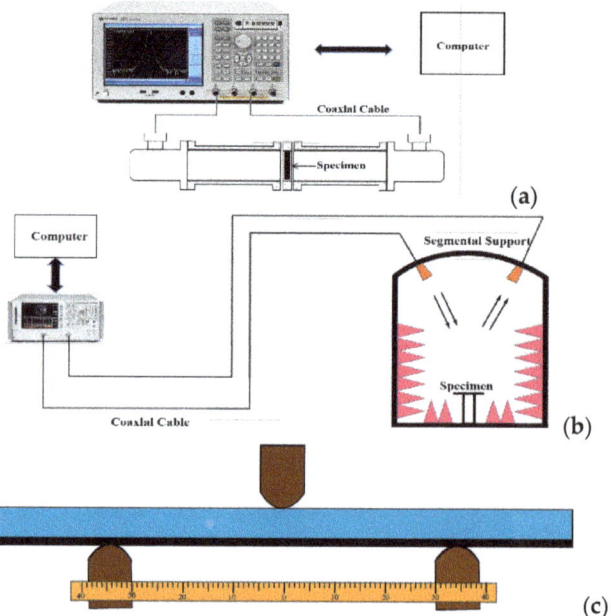

Figure 4. (a) Schematic diagram of the dielectric property test equipment; (b) schematic diagram of the EM reflection loss test equipment; (c) schematic diagram of the three-point bending test.

3. Results and Discussion

3.1. EM Properties

The micro-structure of CIP and CB is shown in Figure 5a,b, respectively. As shown in Figure 5a, the microscopic morphology of CIP was spherical particles with a sphere structure, with an average particle size of about 3.5 μm. Such a spherical structure was also conducive to its uniform distribution in the matrix [16]. As shown in Figure 5b, the microscopic morphology of CB particles was approximately spherical, with a sphere size of about 20–30 nm. Under the electron microscope, the CB particles were agglomerated, caused by the high surface energy of nanoparticles. The agglomerated form made it easy for CB particles to form the conductive path in the matrix, which significantly improved CB's conductivity.

The EM parameters of CB/CIP tested are shown in Figure 6a. From Figure 6a, it can be seen that the real part of the complex dielectric constant of CB/CIP fluctuates very little in the range of 2–18 GHz test frequency, generally distributed in the range of 3.5–3.75, and the imaginary part of the same dielectric constant ranges from 0 to 0.25, showing a weak dielectric loss performance. The real part μ' and imaginary part μ'' of the complex permeability decreased gradually with the increase in frequency and were distributed in the range of 1–1.5 and 0.25–0.5, respectively. Figure 6b shows the curve of the loss tangent of CB/CIP. It can be seen that the dielectric loss tangent of CB/CIP tends to decrease in the measured range and eventually fluctuates around 0. These data indicated that CB/CIP has a weak dielectric loss performance. The magnetic loss tangent of CB/CIP decreased at first and then increased, which indicated that CB/CIP has good magnetic loss performance and is an excellent magnetic loss absorbent.

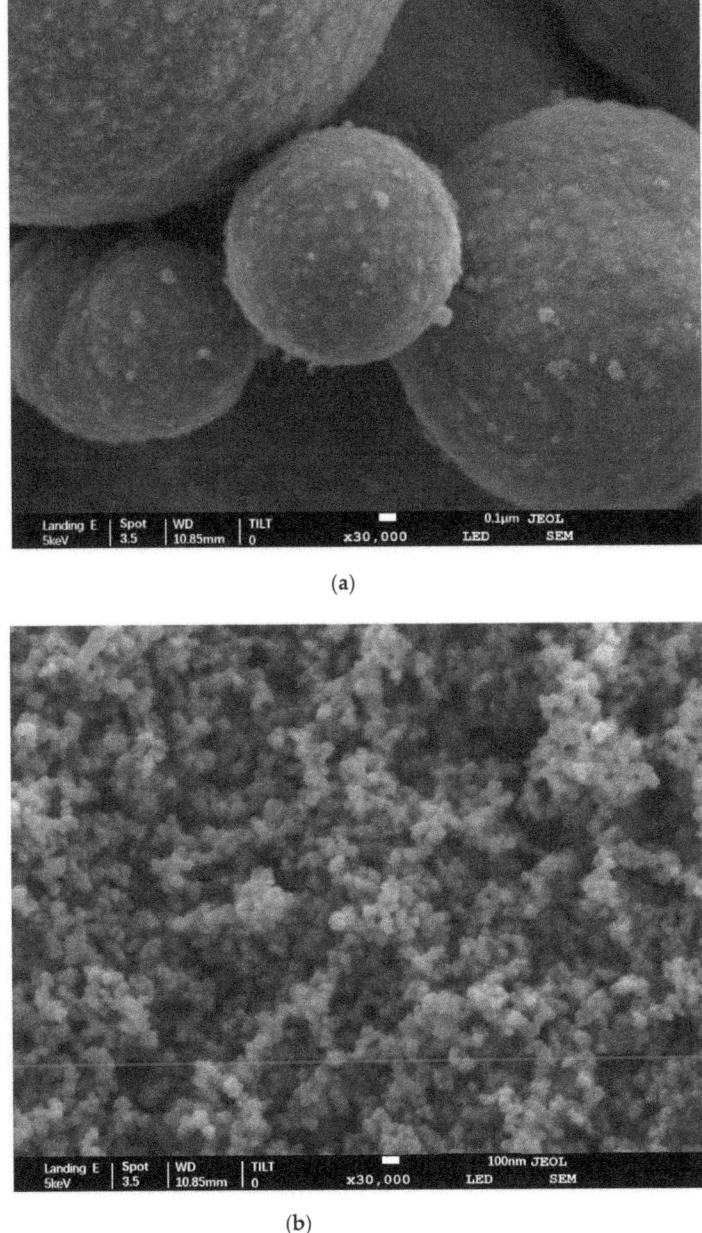

Figure 5. (a) Scanning electron microscope image of CIP; (b) scanning electron microscope image of CB.

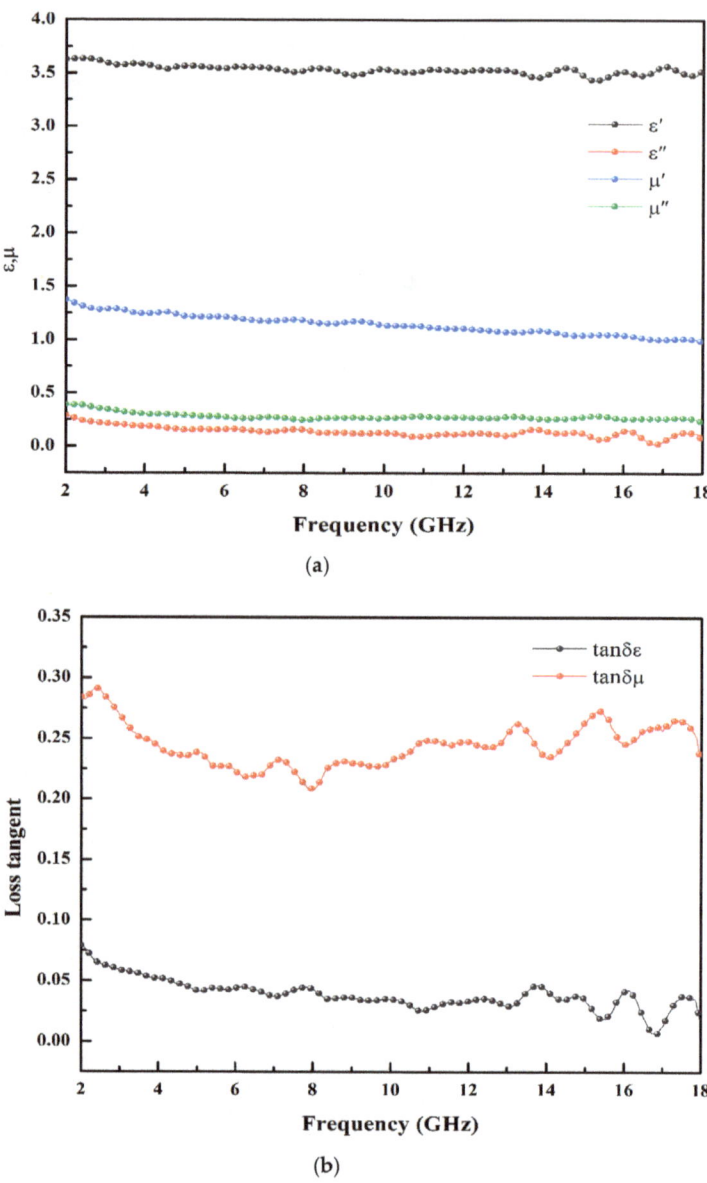

Figure 6. (a) EM parameters of CB/CIP particles; (b) loss tangent of CB/CIP particles.

As shown in Figure 7, the S-band refers to the electromagnetic (EM) wave frequency range between 2 GHz and 4 GHz. The C-band refers to the EM wave segment with a frequency between 4 GHz and 8 GHz. The X-band refers to the EM wave segment with a frequency between 8 GHz and 12 GHz. The Ku-band refers to the EM wave segment with frequencies between 12 GHz and 18 GHz. The S0 reached the RL_{min} of −11.4 dB at a Ku-band (15.66 GHz). The S1 reached the RL_{min} of −15.06 dB at Ku-band (15.48 GHz). Therefore, we found that CB/CIP can improve the absorbing loss capacity of the whole composite.

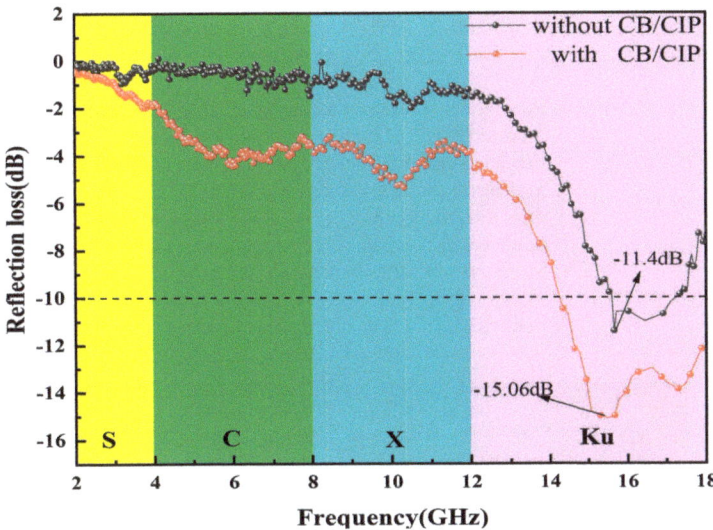

Figure 7. EM wave absorption performance of S0 and S1.

The BF/EP (basalt fiber and epoxy resin) in the 3-D honeycomb woven EM absorbing composite were wave-permeable materials, which have no absorption ability to EM waves. The continuous CF layer mainly reflected the EM wave, and the EM wave that was not lost and was reflected twice after reaching the CF layer to improve the loss efficiency. CB/CIP, as an absorber of EM waves, was distributed in the EP matrix and mainly played the role of magnetic loss. CB/CIP can promote the formation of the absorbing network structure in the matrix, which improved the impedance matching and attenuation loss of the 3-D honeycomb woven EM absorbing composite, and the absorbing loss ability of the whole composite was enhanced. Compared to the two curves, we also found that with the addition of CB/CIP, the EM absorption intensity of the 3-D woven honeycomb EM absorbing composites increases, and the peak value of RL_{min} moved to the low-frequency direction. The RL curve showed the phenomenon that the absorption peak matches the frequency to move to the low frequency, which can be shown through the quarter wavelength theoretical explanation [17]:

$$f_m = \frac{nc}{4d\sqrt{\varepsilon_r \mu_r}} (n = 1, 3, 5 \ldots) \tag{1}$$

f_m: the absorption peaks match frequencies (GHz);
c: the speed of light (m/s);
d: the thickness of the 3-D woven honeycomb EM absorbing composite (mm);
ε_r: the relative dielectric constant;
μ_r: the magnetic permeability.

It can be concluded that the matching frequency of the 3-D woven honeycomb EM absorbing composite was related to its thickness and electromagnetic parameters. According to the effective medium theory [18], it can be seen that the equivalent electromagnetic parameters of the mixture of absorbent and resin were related to the content of absorbent. When the thickness of the 3-D woven honeycomb EM absorbing composite was constant, the increase in CB/CIP content in Equation (1) led to the increase in equivalent EM parameters of absorbent honeycomb, and the corresponding matching frequency of absorption peak decreased; in the final step, the absorption peak moved to low frequency.

The RL curve of the 3-D honeycomb woven EM absorbing composites is shown in Figure 8. The RL_{min} of S1 was -15.06 dB (Figure 8a,b), and the EBA of the Ku–band was

3.66 GHz (Figure 8a,c,d), and the corresponding absorption band accounts for 61%. For S2, RL_{min} was -23.5 dB (Figure 8a,b). Although the EBA of the C–band (Figure 8a,c,d) was only 0.5 GHz and the corresponding absorption band occupies 12.5%, the EBA of the X-band was 2.52 GHz, and the corresponding absorption band occupies 63.8%. S3 obtained an excellent absorption intensity of -25.7 dB. In particular, the EAB of C–band (4–8 GHz) and X-band (8–12 GHz) low-frequency bands were 3.26 GHz and 0.19 GHz, respectively, and the corresponding absorption band occupancy rates were 81.5% and 19%, respectively, indicating that the C-band has excellent EM absorption performance. By comparing the three curves in Figure 8a, we also found that the EM absorption intensity of the 3-D woven honeycomb composites increases, and RL curves' peak value moves to the low-frequency direction with the increase in thickness. The quarter wavelength theory can explain it. According to Equation (1), the thickness was inversely proportional to the absorbing frequency. That is to say, the thicker the thickness, the lower the absorbing frequency.

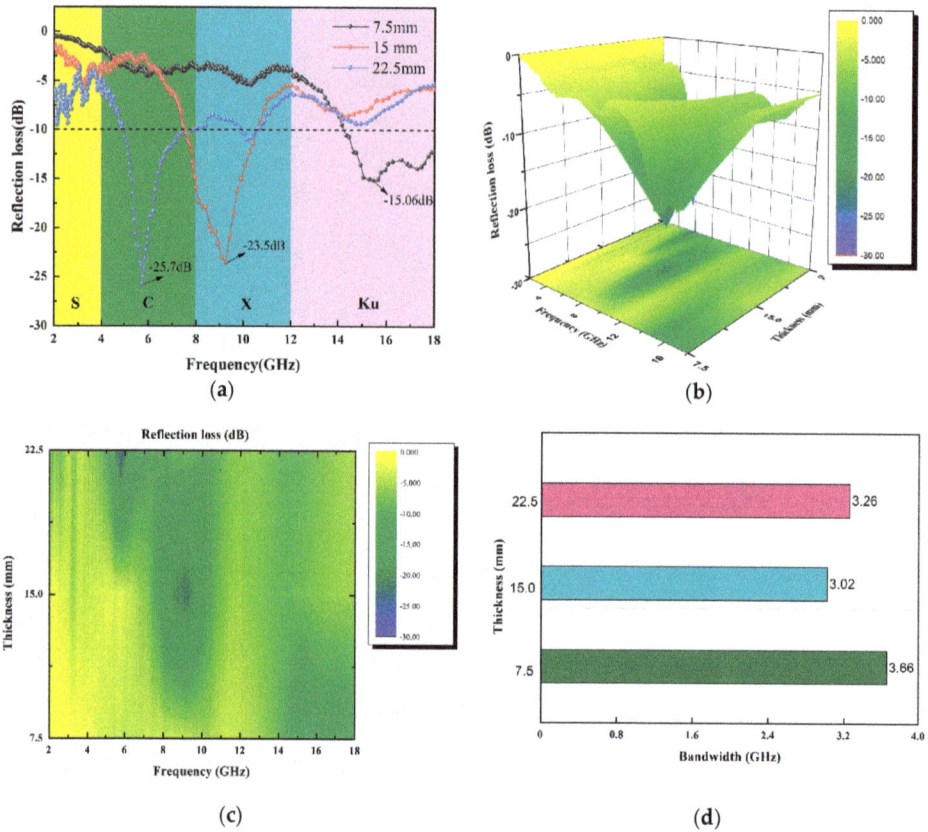

Figure 8. (a) RL curves of S1, S2, S3; (b) 3-D RL of S1, S2, S3; (c) two-dimensional plane RL of S1, S2, S3; (d) EAB of S1, S2, S3.

3.2. Bending Properties

The three-point bending load–displacement curves of the 3-D honeycomb woven EM absorbing composites are shown in Figure 9. The maximum bending load of S0 was 1002.03 N. The maximum bending load of S1 was 846.38 N. The maximum bending load of S2 was 2271.07 N. The maximum bending load of S3 was 2843.03 N. The reason for the maximum bending load decrease in S1 was that when the composites was subjected to bending load, CB/CIP agglomerated in the resin matrix, and a large number of microcracks gathered around the agglomerated particles, which could easily lead to macroscopic cracking caused by stress concentration. The maximum bending load increased with the thickness due to the more fiber bearing force. Therefore, the maximum bending load increased. Figure 9 shows that the curve was divided into three stages. In the initial loading stage, the material as a whole bears the bending load and performs well within the initial displacement, and the bending load–displacement curve also increases linearly. Therefore, the composite material has good linear elastic properties. With the increase in displacement, the composites bore more extrusion pressure. The shape variables were different between the fiber and the resin, the upper fiber bundle was subjected to the extrusion, while the lower fiber bundle was subjected to stretching so that the fiber in the composite material was pulled out and then destroyed the interface between the fiber and matrix. In this way, the modulus and flexural stiffness of the composite decreased, and the matrix cracked. Finally, the bending load–displacement curve showed a peak load and then began to decline. The load–displacement curve was convex on the whole.

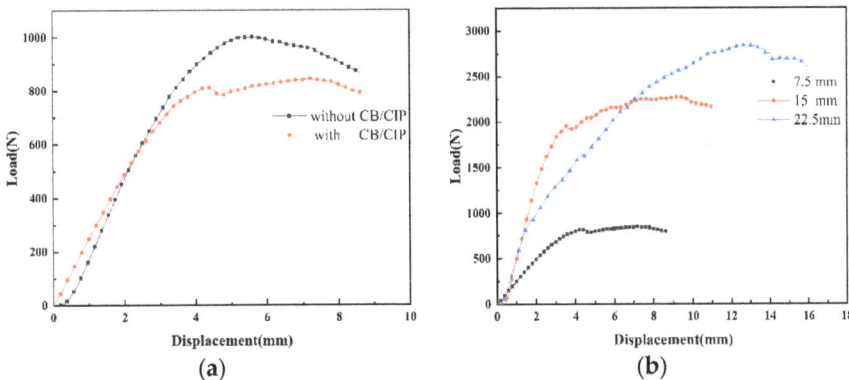

Figure 9. (a) Load–displacement curves of S0 and S1; (b) load–displacement curves of S1, S2 and S3.

The bending strength can be calculated according to the width, thickness, span, and displacement of the sample, and its equation is as follows:

$$\sigma = \frac{3PL}{2bd^2} \qquad (2)$$

σ: the bending strength (MPa);
P: the load (N);
L: the span (mm);
b: the width (mm);
d: the thickness (mm);

When the thickness was 7.5 mm, without CB/CIP, the bending strength was 106.88 MPa. When the thickness was 7.5 mm, with CB/CIP, the bending strength was 90.28 MPa. When the thickness was 15 mm, the bending strength was 121 MPa. When the thickness was 22.5 mm, the bending strength was 101.09 MPa. However, by comparing the load–displacement curve and bending strength of S0 and S1, it was found that the maximum load and bending strength of S1 were reduced. This may be due to the mutual attraction of the CB/CIP surface, resulting in the particle agglomeration phenomenon, resulting in the spatial barrier effect, and CB/CIP will also lead to the decrease in the crosslinking density of EP, resulting in the increase in defects, and the continuity of the EP matrix will be broken. As a result, the strength of the interface between the EP and the fibers weakens, eventually leading to direct cracking and failure of the composites at the CB/CIP aggregation sites. By comparing the bending strength of S1, S2, and S3, it was found that the bending strength of 3-D woven honeycomb EM absorbing composites increased first and then decreased with the increase in layers. This was mainly because, in the bending test, the upper BF was subjected to the action of compression force, the middle layer was subjected to the action of shear force, the lower CF was subjected to the action of tensile force. When the number of BF layers was two, the volume content of BF did not change much because the number of BF layers did not increase much. The bending strength of S2 was eventually increased. Subsequently, the number of layers of BF continued to increase, and the thickness of the 3-D woven honeycomb EM absorbing composites also increased. According to Equation (2), the bending strength was inversely proportional to the square of the thickness and positively proportional to the first power of the span. In contrast, the effect of thickness on bending strength was more obvious. Thus, S3 showed a decrease in bending strength. Additionally, with the increase in the number of BF layers, the volume content of BF gradually increased. Therefore, the effect of the strength of BF was also more obvious. As the strength of BF was lower than that of CF, the bending strength of S3 decreased. The results of the 3-D honeycomb woven EM absorbing composites with different thicknesses were similar. Thus, only the 3-D honeycomb woven EM absorbing composite with S2 was used as an example. Figure 10a–d shows the overall view, partial side view, partial bottom view and partial top view of S2, respectively. It can be seen from Figure 10 that the bending failure modes had a brittle fracture of the fiber bundle, cracking of the matrix, and a typical shear failure. However, the 3-D honeycomb woven EM absorbing composite had good integrity, and there was no delamination.

Figure 10. (a) Overall view of S2; (b) partial side view of S2; (c) partial bottom view of S2; (d) partial top view of S2.

3.3. Mechanical Properties and EM Properties Match

It was necessary to evaluate the comprehensive properties of the composite materials, which met the requirements of the mechanical properties and met the requirements of the absorbing properties. Table 2 shows the mechanical properties and absorbing properties of the 3-D honeycomb woven EM absorbing composites.

Table 2. The mechanical properties and absorbing properties of the 3-D honeycomb woven EM absorbing composites.

Specimen	Mechanical Property	EM Properties	
	Bending Strength (MPa)	RL_{min} (dB)	EAB (GHz)
S0	106.88	−11.4	0.3
S1	90.28	−15.06	3.66
S2	121	−23.5	3.02
S3	101.09	−25.7	3.26

To more intuitively represent the variation trend and amplitude of different thicknesses, mechanical properties, and EM absorbing properties and quickly match the optimal thickness, the concept of the normalized factor β was introduced [19]. Pure BF/CF/EP composite absorbing material without adding CB/CIP particles was regarded as the mechanical performance standard, and the minimum requirement of RL value −10 dB was regarded as the absorbing performance standard. The normalization factors of both were shown in Equations (3) and (4). The data in Table 3 were obtained after the normalization of the data in Table 2.

$$\beta_1 = \frac{\sigma_f - \sigma_0}{\sigma_0}(f = 1, 2, 3) \quad (3)$$

$$\beta_2 = \frac{R_f - (-10)}{-10}(f = 1, 2, 3) \quad (4)$$

β_1: Normalization factor of mechanical properties;
σ_f: Mechanical properties of S1, S2 and S3;
σ_0: Mechanical properties of S0;
β_2: Normalization factor of EM absorbing properties;
R_f: EM absorbing properties of S1, S2 and S3.

Table 3. The mechanical properties and absorbing properties of normalized the 3-D honeycomb woven EM absorbing composites.

Specimen	Mechanical Properties	EM Properties	
	Bending Strength β	RLmin β	EAB (GHz)
S0	0	14%	0.3
S1	−15.53%	50.6%	3.66
S2	13.21%	135%	3.02
S3	−5.79%	157%	3.26

It can be seen from Table 3 that the bending strength of S1 and S3 decreased, so S2 conforms to the principle of matching the absorbing property and mechanical property, and this specimen met the requirements of both absorbing property and bearing capacity.

4. Conclusions

To meet the requirements of modern absorbing composites, which are the integration of load-bearing and EM wave absorption, and improve the low mechanical properties and overall performance of traditional honeycomb absorbing composites, the 3-D woven honeycomb EM absorbing composite was designed. To determine the effects of CB/CIP and thickness on the EM and mechanical properties of the 3-D honeycomb EM absorbing composites, the 3-D honeycomb EM absorbing composite without adding CB/CIP was designed. Three kinds of 3-D honeycomb EM absorbing composites with different thicknesses were prepared.

In the aspect of EM absorbing properties, the CB/CIP can promote the formation of the absorbing network structure in the matrix, which improved the impedance matching and attenuation loss of the 3-D honeycomb woven EM absorbing composite, so the absorbing loss ability of the whole composite was enhanced. When the thickness of the 3-D honeycomb EM absorbing composites was 15 mm and 22.5 mm, respectively, the composites' absorptivity in the X-band and C-band was 63.8% and 81.5%, respectively. This was because with the increase in thickness, the fiber layer number was increased, and the hole was increased, the electromagnetic wave occurred in the material reflection, and the number of refractions was increased. Thereby, there was an increase in the electromagnetic waves absorbed. Besides, with CB/CIP and the increase in thickness, the RL peak of 3-D woven honeycomb EM absorbing composites moved to low frequency. This fit the quarter wavelength theory.

In terms of mechanical properties, the EP matrix's continuity was broken easily due to the large amount of agglomeration of CB/CIP, and the strength between the resin and dimensional interface was weakened, thus reducing the mechanical properties of the 3-D honeycomb EM absorbing composites. Additionally, the results showed that the bending load of the 3-D honeycomb EM absorbing composites increased with the increase in the thickness of the composites, and the bending strength increased first and then decreased with the increase in the thickness of the composites. The 3-D honeycomb EM absorbing composites' failure modes were mainly fiber bundle brittle fracture, matrix cracking, and typical shear failure. During the experiment, the 3-D honeycomb EM absorbing composite had no delamination and splitting phenomenon, and the overall performance was good.

After normalizing the mechanical and absorbing properties of the 3-D honeycomb EM absorbing composites, the optimal load/absorbing ratio was S2. The thickness was 15 mm, the maximum bending load was 2271.07 N, the bending strength was 121 MPa, the RL_{min} was -23.5 dB, and the EAB was 3.02 GHz; the composites satisfied with both load-bearing and EM wave absorption capacity.

This work provides a simple way to improve the mechanical properties and EM absorption capacity of structural EM absorbing composites with the 3-D woven honeycomb structure. However, we only investigated the EM absorbing property of the composites in regard to adding absorbent and different thicknesses, and we need to design further a 3-D woven honeycomb composite with different aperture sizes to achieve EM absorbing so that 3-D woven honeycomb composite can be used in weapon systems, such as fighters, warships, and missiles.

Author Contributions: L.-H.L. and W.-D.L. designed the experiments. W.-D.L. wrote the manuscript and performed the experiments and analyzed the data. B.-Z.S. contributed to discussion. All authors have read and agreed to the published version of the manuscript.

Funding: This research was funded by the Key Laboratory of High Performance Fibers and Products, Ministry of Education, Donghua University, Shanghai, China and the National Science Foundation of Liaoning Province (2019-MS-017).

Institutional Review Board Statement: Not applicable.

Informed Consent Statement: Not applicable.

Data Availability Statement: The data presented in this study are available on request from the corresponding author.

Conflicts of Interest: The authors declare no conflict of interest.

References

1. Luo, H.; Chen, F.; Wang, X.; Dai, W.Y.; Xiong, Y.; Yang, J.J.; Gong, R. A novel two-layer honeycomb sandwich structure absorber with high-performance microwave absorption. *Compos. Part A* **2019**, *119*, 1–9. [CrossRef]
2. Liu, W.D.; Lyu, L.H. Research Progress of Three-dimensional Woven Wave-absorbing Composite Materials. *Cotton Text. Technol.* **2020**, *48*, 81–84.
3. Guo, X.S.; Chen, L.; Sun, H.M.; Gu, Z.Z. Design and development of irregular hexagonal honeycomb absorbing materials. *Compos. Sci. Eng.* **2018**, *12*, 67–71.
4. Kim, P.C.; Lee, D.G. Composite sandwich constructions for absorbing the electromagnetic waves. *Compos. Struct.* **2009**, *87*, 161–167. [CrossRef]
5. Li, W.C.; Xu, L.Y.; Zhang, X.; Gong, Y.; Ying, Y.; Yu, J.; Zheng, J.; Qiao, L.; Che, S. Investigating the effect of honeycomb structure composite on microwave absorption properties. *Compos. Commun.* **2020**, *19*, 182–188. [CrossRef]
6. Wang, H.; Xiu, X.; Wang, Y.; Xue, Q.; Ju, W.B.; Che, W.Q.; Liao, S.; Jiang, H.; Tang, M.; Long, J.; et al. Paper-based composites as a dual-functional material for ultralight broadband radar absorbing honeycomb. *Compos. Part B* **2020**, *202*, 1–9. [CrossRef]
7. Sun, P.C.; Wang, L.M.; Wang, T.; Huang, J.; Sha, X.W.; Chen, W. Structural optimization of the design of a double-layer absorbing honeycomb composite. *J. Beijing Univ. Chem. Technol.* **2019**, *46*, 58–64.
8. Menta, V.G.K.; Vuppalapati, R.R.; Chandrashekhara, K. Manufacturing and mechanical performance evaluation of resin-infused honeycomb composites. *J. Reinf. Plast. Compos.* **2012**, *31*, 415–423. [CrossRef]
9. Ji, G.F.; Ouyang, Z.Y.; Li, G.Q. Debonding and impact tolerant sandwich panel with hybrid foam core. *Compos. Struct.* **2013**, *103*, 143–150. [CrossRef]
10. Xiong, J.; Zhang, M.; Stocchi, A. Mechanical behaviors of carbon fiber composite sandwich columns with three dimensional honeycomb cores under in-plane compression. *Compos. Part B* **2014**, *60*, 350–358. [CrossRef]
11. Wei, F.; Li, D.D.; Li, J.L.; Li, J.Z.; Yuan, L.J.; Xue, L.L.; Sun, R.-J.; Meng, J.-G. Electromagnetic properties of three-dimensional woven carbon fiber fabric/epoxy composite. *Text. Res. J.* **2018**, *88*, 2353–2361.
12. Xiao, B.Q.; Huang, Q.W.; Chen, H.X.; Chen, X.B.; Long, G.B. A fractal model for capillary flow through a single tortuous capillary with roughened surfaces in fibrous porous media. *Fractals* **2021**, *29*, 2150017. [CrossRef]

13. Bayraktar, G.B.; Kianoosh, A.; Bilen, D. Fabrication of Woven Honeycomb Structures for Advanced Composites. *Text. Leather Res.* **2018**, *1*, 114–119. [CrossRef]
14. Lv, L.H.; Huang, Y.L.; Cui, J.R. Bending Properties of Three-dimensional Honeycomb Sandwich Structure Composites: Experiment and Finite Element Method Simulation. *Text. Res. J.* **2018**, *88*, 2024–2031. [CrossRef]
15. Zahid, B.; Jamshaid, H.; Rajput, A.W. Effect of Cell Size on Tensile Strength and Elongation Properties of Honeycomb Weave. *Ind. Text.* **2019**, *70*, 133–138. [CrossRef]
16. Liu, L.; Duan, Y.; Ma, L. Microwave absorption properties of a wave-absorbing coating employing carbonyl-iron powder and carbon black. *Appl. Surf. Sci.* **2010**, *257*, 842–846. [CrossRef]
17. Dai, X.H. Research on Electromagnetic Wave Absorption Performance of Honeycomb Structure Composites. Master's Thesis, Dalian University of Technology, Dalian, China, June 2019.
18. Matttsine, S.M.; Hock, K.M.; Liu, L. Shift of resonance frequency of long conducting fibers embedded in a composite. *J. Appl. Phys.* **2003**, *94*, 1146–1154. [CrossRef]
19. Zhang, X.F. Study on Preparation and Properties of Structural Wave Absorbing Composite Materials via VARI Process. Master's Thesis, Nanjing University of Aeronautics and Astronautics, Nanjing, China, March 2019.

Article

Dyeing Para-Aramid Textiles Pretreated with Soybean Oil and Nonthermal Plasma Using Cationic Dye

Mary Morris [1], Xiaofei Philip Ye [1,*] and Christopher J. Doona [2,3]

[1] Department of Biosystems Engineering and Soil Science, The University of Tennessee, Knoxville, TN 37996, USA; maramorr@vols.utk.edu
[2] U.S. Army Combat Capabilities Development Command—Soldier Center, Natick, MA 01760, USA; doonac@mit.edu
[3] Research Affiliate, Massachusetts Institute of Technology—Institute for Soldier Nanotechnologies, 77 Massachusetts Ave., NE47-4F, Cambridge, MA 02139, USA
* Correspondence: xye2@utk.edu; Tel.: +1-(865)-974-7129

Citation: Morris, M.; Ye, X.P.; Doona, C.J. Dyeing Para-Aramid Textiles Pretreated with Soybean Oil and Nonthermal Plasma Using Cationic Dye. *Polymers* **2021**, *13*, 1492. https://doi.org/10.3390/polym13091492

Academic Editors: Tarek M. Abou Elmaaty and Maria Rosaria Plutino

Received: 18 April 2021
Accepted: 3 May 2021
Published: 6 May 2021

Publisher's Note: MDPI stays neutral with regard to jurisdictional claims in published maps and institutional affiliations.

Copyright: © 2021 by the authors. Licensee MDPI, Basel, Switzerland. This article is an open access article distributed under the terms and conditions of the Creative Commons Attribution (CC BY) license (https://creativecommons.org/licenses/by/4.0/).

Abstract: The increasing use of functional aramids in a wide array of applications and the inert nature of aramids against conventional dye and print methods requires developing new dyeing methods. This study aims to use environmentally friendly method with a cationic dye as an alternative for dyeing para-aramid fabrics. Experiments used a multi-factorial design with functions of pretreatment, dye solvent (water and/or glycerol) and auxiliary chemical additives (swelling agent and surfactant) and a sequential experimentation methodology. The most effective dyeing procedures involved the following steps: (i) pretreatments of the fabrics with soybean oil and nonthermal plasma (NTP), (ii) using water at T = 100 °C as the dye solvent, and (iii) omitting other chemical additives. With a commercial cationic dye, these conditions achieved a color strength in K/S value of 2.28, compared to ~1 for untreated samples. FTIR analysis revealed that a functional network formed on the fibers and yarns of the fabrics by chemical reactions of excited plasma species with double bonds in the soybean oil molecules was responsible for significantly improving the color strength. These results extend the potential uses of a renewable material (soybean oil) and an environmentally friendly technology (NTP) to improve the dyeing of para-aramid textiles and reduce the use of harsh dye chemicals.

Keywords: soybean oil; glycerol; nonthermal plasma; para-aramid textiles; cationic dye

1. Introduction

Aramid (portmanteau combining aromatic and polyamide) fibers are high-performance fibers with properties of a high tensile strength, high melting point, chemical resistance to a range of organic solvents and exceptional flame resistance [1]. Consequently, aramid fibers have found wide applications in various technical areas, including textiles for apparel such as body armor (bullet-proof vests and helmets), puncture–resistant correctional wear, fire-protective clothing and sportswear and other applications such as brake pads, gaskets, hot-air filters, industrial belts and ropes, reinforced composites, tire cords and the strength member in fiber optics.

There are two main types of aramids: meta-aramids and para-aramids, and these two groups have different properties due to differences in their molecular structures. At the fundamental level, para-aramid fibers consist of poly(p-phenylene terephthalamide) molecules estimated to be 230 nm long, with stiff para-linked aromatic rings and densely arranged hydrogen bond donors and acceptors throughout their backbones. This inherent molecular rigidity, combined with strong intermolecular hydrogen bonding interactions, enables the molecules to achieve excellent alignment with their neighbors, resulting in a highly anisotropic unit cell consisting of covalent bonds, hydrogen bonds and van der Waals interactions along each fundamental axis, forming a highly crystalline structure [1]. The building-block molecules of meta-aramid are poly(m-phenylene isophthalamide) that bind via meta-linked aromatic rings to result in a semi-crystalline fiber with the molecular chain oriented along the fiber axis.

With the increasing array of applications using aramids, there is also the need to find alternative methods for dyeing aramid textiles. It has been a significant challenge to dye or print aramid fabrics to a high color strength using conventional dyeing and printing methods, especially for continuous filament para-aramids due to their highly crystalline structure and chemical inertness. Currently, the color of woven aramid materials uses primarily solution dyeing methods, in which the coloring of the yarns in the woven or knit fabric is determined by adding colorant to the polymer dope at the time the aramid filament is produced, thereby limiting the color options for the fabrics and their use in potential new applications. Various surface modification methods have been attempted to improve the dyeing of aramids, including chemical treatments with strong acids [2,3] and auxiliary additives [4], physical approaches using UV/O_3 irradiation [5] or nonthermal plasma [6], and chemical grafting using poly(acrylic acid) [7] or a diblock copolymer derived from methacryloyloxy-ethyl-trimethylammonium chloride [8], which claimed improved dyeability with different types of dyes. Most of these methods focused on the surface modification of meta-aramids, and few reported successes in dyeing para-aramids [9–11].

In terms of the dye characteristics suited for dyeing aramids, disperse dyes were developed to dye synthetic fibers and worked well on nylon [12], which were aliphatic or semi-aromatic polyamides, but limited success had been reported in dyeing aramids without harsh chemical treatments [4]. In solution dyeing, the pigments are added to the sulfuric acid solution just before spinning to achieve coloration of the resultant para-aramid fibers [13]. Further, disperse dyes are so-called because they tend to have low solubility in water and require special dispersing agents to effectuate coloration. Another type of dye, the cationic dyes, are exclusively used in solution dyeing, with the help of special dyeing auxiliaries and/or solvents in the case of meta-aramid fibers. In contrast, cationic dyes can dissociate into positively charged chromophore ions in aqueous solution and interact with the negative groups on the fiber molecules to form salts, which can become firmly attached to the fibers for dyeing. Using conventional dyeing methods to investigate dyeing characteristics of meta-aramid fibers with some commercial dyes, Kim and Choi [14] reported that the cationic dyes showed comparatively higher exhaustion yield comparing to those of disperse dyes and acid dyes, and under acidic conditions in the range of pH 3 to 5, the stability of cationic dyes could be enhanced, leading to higher adsorption.

We hypothesize that surface modification of para-aramids via nonthermal plasma (NTP) treatment, particularly an oxygen-containing plasma method, would introduce anionic dyeing sites on the surface to facilitate dyeing. Therefore, the objective of this study is to dye para-aramid textiles with a cationic dye using renewable materials and an environmentally friendly technology (NTP). Additionally, since the trend in the textile dyeing industry is to reduce or avoid the use and disposal of environmentally-unfriendly chemical additives [15], glycerol is used in this study as an alternative solvent to water or ethanol due to glycerol's non-toxicity, low volatility and high boiling point (290 °C), all of which make glycerol environmentally safe and suitable for use in the textile dyeing industry [16,17]. Further, glycerol is also used in this study, because there is a current worldwide effort to valorize the excessive amount of crude glycerol that has been generated as a result of massive biodiesel production [18–21].

Textile dyeing processes commonly add surfactants to the dyeing media, to ensure the uniform dispersion of dye in the media that promotes penetration of the dye into the fiber matrix [22]. Swelling agents or dyeing accelerators swell the fibers to facilitate the penetration of dye into the fibers. Swelling agents are particularly useful for increasing the dye substantivity in some instances with highly crystalline synthetic and blended fibers [23,24]. In this study, the nonionic surfactant and emulsifier Polysorbate 80 (TWEEN 80) and the swelling agent benzyl alcohol, both of which have been reported to have benefits in dyeing aramids [25,26], were tested as additives to improve dyeing.

2. Materials and Methods

2.1. Materials

The textiles tested in these experiments consisted of continuous filament 300 Denier para-aramid yarns in a tightly woven plain weave construction, as provided by the U.S. Army Combat Capabilities Development Command—Soldier Center (Natick, MA, USA). The cationic dye Basic Blue 11 (Victoria Blue R, CAS Number 2185-86-6) was purchased from Sigma-Aldrich (St. Louis, MO, USA); glycerol (99.5% purity) and acetic acid (glacial) were purchased from Fisher Scientific (Waltham, MA, USA); Polysorbate 80 (TWEEN 80) and benzyl alcohol were purchased from Chem Center @ Amazon.com; and food grade refined soybean oil (100 g contains approximately 16 g of saturated fat, 23 g of monounsaturated fat, and 58 g of polyunsaturated fat) and laundry detergent (ECOS® Plus with Stain-Fighting Enzymes) were purchased from a local supermarket. All reagents and materials were used as-is without further purification.

2.2. Experimental Methods

The sequential experimentation methodology used in this study (Experiments A–F) was directed toward the goal of finding the combination of experimental factors that achieve the highest color strength of para-aramid fabrics dyed with a cationic dye. The experimental design used sequential factorial designs to quickly identify critical factors and conditions that improved dyeing. The factorial functions consisted of: (i) pretreatment with acetic acid, soybean oil and/or NTP treatment, (ii) dye solvent using water versus glycerol and (iii) the addition of the auxiliary additives of swelling agent and surfactant. The experimental design and subsequent analysis were carried out using specially designed software Design-Expert 6.0 (Stat-Ease, Inc., Minneapolis, MN, USA).

2.3. Pretreatment (Acetic Acid, Soybean Oil, NTP)

Para-aramid textile fabric samples (~645 mm^2 or ~1 in^2) were pretreated by simply immersing the fabric samples into a Petri dish containing either 20 wt.% aqueous acetic acid solution or soybean oil, according to an experimental design. The samples were removed after 15 h, transferred onto a paper towel and hand-pressed with a roller to force out any excess liquid.

For fabric samples subjected to NTP exposure, the NTP treatment was carried out using an in-house made Surface Dielectric Barrier Discharge (SDBD) system. The SDBD method was chosen because it generates a higher density of micro-discharges that are limited to the sample surface compared to the Volume Dielectric Barrier Discharge (VDBD) method. The SDBD therefore avoids the formation of pinholes in the para-aramid fabric samples, which is commonly associated with VDBD and caused by hot electron bombardment of the para-aramid fabrics. Teflon-coated aramids have been made wettable after a 30 s exposure to this SDBD.

The SDBD system has been described previously for dyeing with a disperse dye [27] and is summarized here. The SDBD apparatus uses ambient air as the feed gas and comprises two electrodes that are separated by an alumina dielectric plate (dimensions = 108 × 95 × 1 mm^3 thick) and powered by a high voltage power source with a sinusoidal high voltage of 9.2 kV that was tuned to a 23.2 kHz resonance frequency (see Figure 1). Embedded at the top of the alumina plate was an induction electrode made of rectangular copper tape, and at the bottom of the alumina plate was a discharge electrode (17 tungsten strips interconnected). Fabric samples were placed on the rotating stage and the SDBD plate was lowered to a height 1 mm above the sample. When the power was switched on, the plasma emissions, which included reactive oxygen and nitrogen species (RONS) and other high-energy radicals, interacted with the fabric samples (typically for 1–2.5 min). Then, the power was turned off and the sample was removed for dyeing.

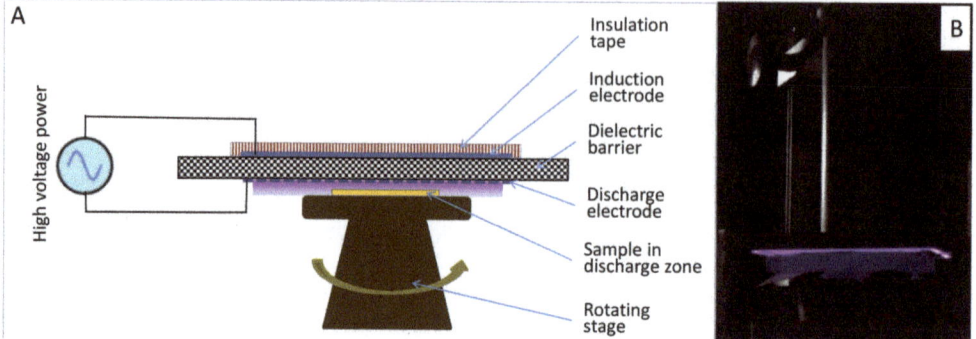

Figure 1. (**A**) Diagram of the NTP system and (**B**) photo showing a sample under glowing plasma discharge.

2.4. Dyeing Experimental Procedures

The cationic dye Victoria Blue R is soluble in water and in glycerol, so distilled water, glycerol and aqueous glycerol solutions were tested and compared as the dye solvent. The aqueous glycerol solutions are expressed as their volumetric percent of glycerol. The dye solvents were moderately heated (T ≈ 50 °C) with vigorous stirring using a magnetic stirrer, and appropriate amounts of Victoria Blue R dye, acetic acid, surfactant and swelling agent were added to the respective solvents in accordance with the design of the experiment. The dye bath liquid was dispensed into separate vials that were simultaneously heated and stirred on a Reacti-Therm system (ThermoFisher Scientific, Waltham, MA, USA). Fabric samples were loaded into the vials for dyeing. The dye bath and fabric samples took about 30 min to reach a designated temperature, held at the temperature for 1 h, then cooled for 20 min with the Reacti-Therm system shut off. After the dyeing, samples were rinsed in flowing warm tap water for 2 min, rinsed in cold water for 2 min, then dried in a programmable convective oven at T = 150 °C for 2 min (after starting at T = 30 °C and increasing the temperature at a rate of 30 °C/min) to fix the dye.

2.5. Color Strength Analysis

Prior to determining the color strength, the fabric samples were washed with laundry detergent using a homemade tumbler, in accordance with the ISO standard 105-C10:2006 protocol [28], then rinsed and dried as described in Section 2.4 above.

Estimates of the color strength for each of the dyed fabric samples were determined at the wavelength of maximum absorbance for Victoria Blue R dye (λ_{max} = 615 nm) by measuring percent spectral reflectance (%R) in the visible range with a SPECTRO 1 spectrophotometer (Variable Inc., Chattanooga, TN, USA). Results were expressed as the samples' absorption (K) and scattering (S) characteristics as its K/S value, which varies approximately linearly with colorant concentration, according to the Kubelka-Monk equation (Equation (1)) [29].

$$K/S = \frac{(1 - 0.01R)^2}{2(0.01R)} \quad (1)$$

2.6. FTIR Analysis

In order to monitor some of the chemical changes occurring during the pretreatment and dyeing processes, Attenuated Total Reflection—Fourier Transform Infrared (ATR-FTIR) spectroscopy was used to analyze the fabric samples at different stages of the pretreatment and dyeing processes with the Victoria Blue R dye. In summary, fabric samples were placed on a potassium bromide (KBr) plate, pressed under the germanium crystal of ATR (UMA 400, Varian Inc., Palo Alto, CA, USA) and scanned in the mid-IR region (500–4000 cm^{-1} with a 4 cm^{-1} resolution) with an FTIR spectrometer (Excalibur 3100, Varian Inc., Palo Alto,

CA, USA) equipped with an Attenuated Total Reflection (ATR) accessory. The ATR-FTIR spectra were displayed in absorbance units with each spectrum representing an average of 128 scans and taking into account the background spectrum acquired using a blank KBr plate.

3. Results and Discussion

3.1. Experiments A–F: Analysis

Experiment A was a 4-factor, 2-level, full factorial experiment designed to probe the effects of dye solvent (water or glycerol), pretreatment with acetic acid and/or NTP treatment for 60 sec and the addition of the swelling agent benzyl alcohol. The dyeing results are also shown in Table 1.

Table 1. Experiment A: experimental design and results [@].

Sample	Dye Solvent	Pretreatment	NTP Time (s)	Swelling Agent	K/S	K/S Std *
A1	Water	None	60	None	1.26	0.009
A2	Water	None	0	None	1.20	0.054
A3	Water	Acetic acid	60	None	1.22	0.011
A4	Water	Acetic acid	0	None	1.12	0.018
A5	Water	None	60	Benzyl–OH	1.44	0.003
A6	Water	None	0	Benzyl–OH	1.56	0.026
A7	Water	Acetic acid	60	Benzyl–OH	1.74	0.021
A8	Water	Acetic acid	0	Benzyl–OH	1.59	0.016
A9	Glycerol	None	60	None	0.42	0.005
A10	Glycerol	None	0	None	0.39	0.009
A11	Glycerol	Acetic acid	60	None	0.45	0.004
A12	Glycerol	Acetic acid	0	None	0.40	0.005
A13	Glycerol	None	60	Benzyl–OH	0.41	0.002
A14	Glycerol	None	0	Benzyl–OH	0.42	0.003
A15	Glycerol	Acetic acid	60	Benzyl–OH	0.48	0.019
A16	Glycerol	Acetic acid	0	Benzyl–OH	0.47	0.004

[@] Experiment conducted at soaking time in 20 wt.% acetic acid solution = 15 h, dye concentration = 0.1 wt.%, Benzyl alcohol concentration = 2 wt.%, T = 140 °C, dyeing time = 1 h. * Values represent mean and standard error of 3 replicates.

Overall, the highest K/S value achieved in Experiment A is only 1.74 (sample A7) through the combination of water as solvent, pretreatments with acetic acid and NTP (60 s) and the addition of swelling agent benzyl alcohol. The Analysis of Variance (ANOVA) factorial model terms were selected based on the half-normal probability plot, which indicated that all the factors and some of their interactions were significant (t-tests for coefficients, $p < 0.02$). Since all of the factors are involved in interactions, statistical interpretations of only the significant interaction terms are shown in Figure 2.

Evidently in Figure 2, the dye solvent had the most significant influence on the color strength (K/S value), with water performing significantly better than glycerol. Adding benzyl alcohol to the dye bath increased the color strength, and the effect of benzyl alcohol was slightly enhanced by the acetic acid pretreatment with water as the dye solvent. The effects of pretreatment with acetic acid and NTP were not obvious, slightly improving the color strength with their synergy.

A significant loss of water was observed in the dye bath during Experiment A due to the evaporation of water at the dyeing temperature T = 140 °C. The use of glycerol as the dye solvent effectuated dyeing at T = 140 °C without the concomitant evaporative loss of solvent from the dye bath but reached significantly lower values of color strengths. The cationic dye might not be well-dissociated in glycerol as the solvent, and the high viscosity of pure glycerol compared to water and the occurrence of hydrogen bonding between the dye and glycerol might hinder the mobility of the cationic dye and retard its diffusion onto the para-aramids.

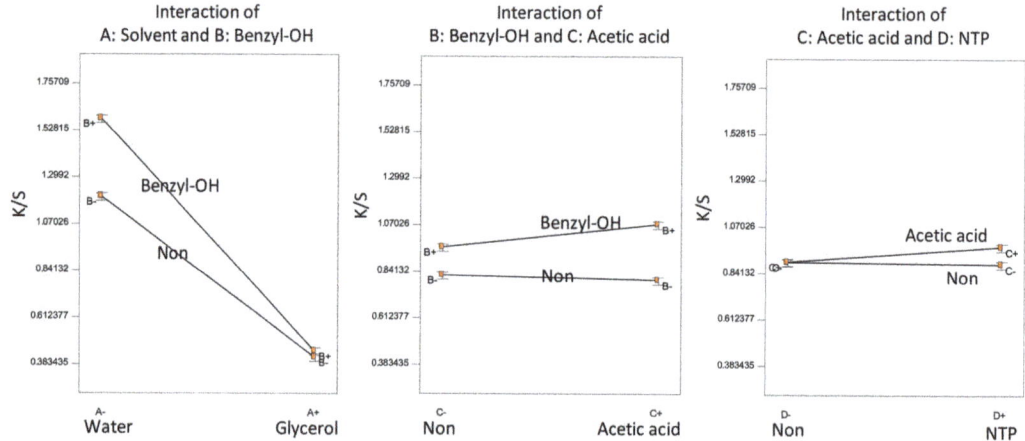

Figure 2. Statistical inference of Experiment A: significant factorial interactions.

To overcome the drawbacks of using glycerol as the dye solvent, full factorial Experiment B was designed using 50% or 80% aqueous glycerol solutions as the dye solvent. Additional factors were investigated and included a pretreatment of soaking the para-aramid fabrics in soybean oil and/or applying further treatment with NTP and adding benzyl alcohol as the swelling agent. Table 2 lists the detailed experimental design and resultant K/S values. Even though using 50% or 80% aqueous glycerol solutions as the dye solvent suppressed the loss of the dye bath solution, the highest K/S value achieved in this experiment was only 0.94 (sample B11) with 50% aqueous glycerol solution as the solvent and soybean oil and NTP (60 s) as the pretreatment.

Table 2. Experiment B: experimental design and results [@].

Sample	Dye Solvent	Pretreatment	Swelling Agent	NTP Time (s)	K/S	K/S Std *
B1	80% Glycerol	Soy oil	Benzyl–OH	60	0.70	0.015
B2	80% Glycerol	Soy oil	Benzyl–OH	0	0.57	0.012
B3	80% Glycerol	Soy oil	None	60	0.79	0.014
B4	80% Glycerol	Soy oil	None	0	0.56	0.004
B5	80% Glycerol	None	Benzyl–OH	60	0.55	0.003
B6	80% Glycerol	None	Benzyl–OH	0	0.53	0.004
B7	80% Glycerol	None	None	60	0.46	0.017
B8	80% Glycerol	None	None	0	0.44	0.013
B9	50% Glycerol	Soy oil	Benzyl–OH	60	0.84	0.005
B10	50% Glycerol	Soy oil	Benzyl–OH	0	0.67	0.009
B11	50% Glycerol	Soy oil	None	60	0.94	0.001
B12	50% Glycerol	Soy oil	None	0	0.92	0.005
B13	50% Glycerol	None	Benzyl–OH	60	0.81	0.014
B14	50% Glycerol	None	Benzyl–OH	0	0.70	0.010
B15	50% Glycerol	None	None	60	0.72	0.006
B16	50% Glycerol	None	None	0	0.82	0.013

[@] Experiment carried out with soaking in soybean oil time = 15 h, dye concentration = 0.1 wt.%, Benzyl alcohol concentration = 2 wt.%, T = 140 °C and dyeing time = 1 h. * Mean and standard error of 3 replicates.

The analysis of variance (ANOVA) of this factorial design indicates that the main effect of all the factors and some of their interactions are significant ($p < 0.005$). The half-normal probability plot was used to select the ANOVA model terms, and only the significant interaction terms of the model are presented in Figure 3.

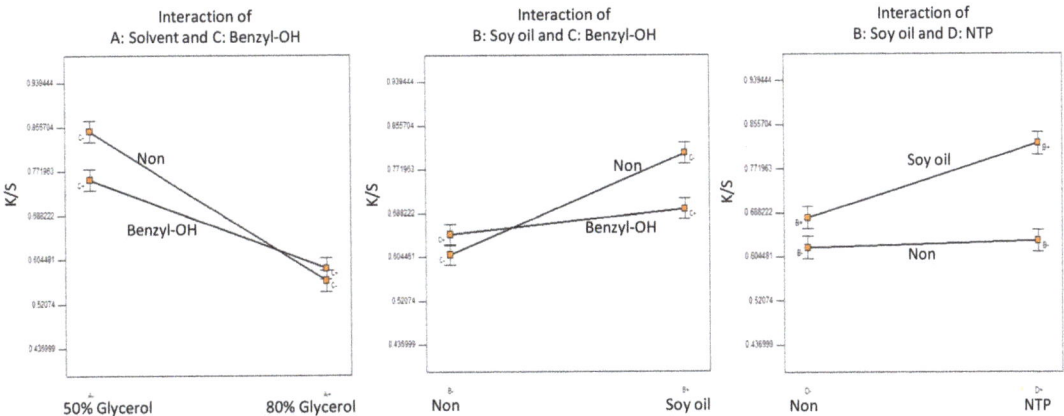

Figure 3. Statistical inference of Experiment B: significant factorial interactions.

The 50% aqueous glycerol solution performed better than the 80% aqueous glycerol solution as the dye solvent (Figure 3). Benzyl alcohol appeared to be incompatible with aqueous glycerol or soybean oil, which might hinder its diffusion and swelling functions and consequently limited the benefits of soybean oil on dyeing. Soaking the para-aramids in soybean oil slightly improved the dyeing color strength, while soaking the para-aramids in soybean oil with subsequent NTP treatment for 60 s increased the color strength significantly. Without a pretreatment of soybean oil pretreatment, NTP had negligible effect on the dyeing strength.

The full factorial design for Experiment C used a lower dyeing temperature (T = 90 °C), compared 50% aqueous glycerol solution with water as the dye solvent and used soybean oil pretreatments and NTP treatment times of 90 s. The experimental design and dyeing results are compiled in Table 3. The highest K/S value achieved in this experiment is 1.72 (sample C11), which is higher than the highest value obtained in Experiment B (0.94). Experiment B was carried out at a higher temperature (T = 140 °C vs. T = 90 °C for Experiment C), used soybean oil with a shorter NTP treatment time (60 s vs. 90 s for Experiment C) and used 50% or 80% aqueous glycerol solutions as the dye solvent (vs. water or 50% aqueous glycerol in Experiment C).

The half-normal probability plot was used to select the ANOVA model terms and resulted in a significant model (F-test, $p < 0.0001$). All of the terms included in the model were significant (t-tests, $p < 0.007$), except for the main effect of swelling agent (t-test, $p = 0.59$). Since all of the factors were involved in interactions, only the significant interaction terms are interpreted in Figure 4.

Table 3. Experiment C: experimental design and the results [@].

Sample	Dye Solvent	Pretreatment	Swelling Agent	NTP Time (s)	K/S	K/S Std *
C1	50% Glycerol	Soy oil	Benzyl–OH	90	0.91	0.021
C2	50% Glycerol	Soy oil	Benzyl–OH	0	0.60	0.002
C3	50% Glycerol	Soy oil	None	90	1.06	0.045
C4	50% Glycerol	Soy oil	None	0	0.57	0.006
C5	50% Glycerol	None	Benzyl–OH	90	0.96	0.009
C6	50% Glycerol	None	Benzyl–OH	0	0.74	0.015
C7	50% Glycerol	None	None	90	1.04	0.027
C8	50% Glycerol	None	None	0	0.66	0.013
C9	Water	Soy oil	Benzyl–OH	90	1.66	0.262
C10	Water	Soy oil	Benzyl–OH	0	1.15	0.008
C11	Water	Soy oil	None	90	1.72	0.072
C12	Water	Soy oil	None	0	0.86	0.008
C13	Water	None	Benzyl–OH	90	1.68	0.024
C14	Water	None	Benzyl–OH	0	1.31	0.104
C15	Water	None	None	90	1.09	0.034
C16	Water	None	None	0	0.92	0.012

[@] Experiment was carried out with soaking in soybean oil for time = 15 h, dye concentration = 0.1 wt.%, Benzyl alcohol concentration = 2 wt.%, T = 90 °C and dyeing time = 1 h. * Mean and standard error of 3 replicates.

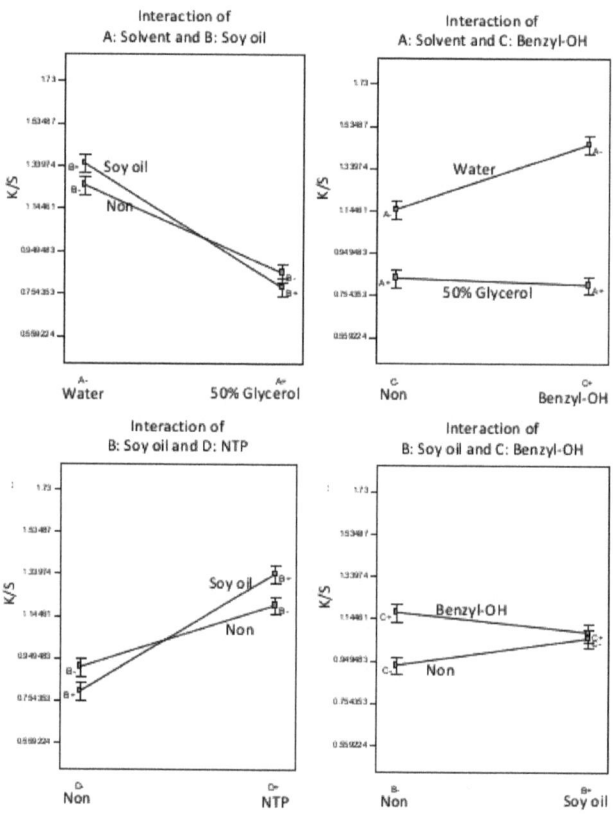

Figure 4. Statistical inference of Experiment C: significant factorial interactions.

According to the half-normal probability plot, the solvent had the most influence on color strength among all of the factors, with water as the solvent performing significantly better than the 50% aqueous glycerol solution. Soybean oil pretreatment followed by NTP significantly improved the color strength, although NTP treatment without the soybean oil pretreatment also improved the color strength, but to a lesser degree. Benzyl alcohol performed better in water than in 50% aqueous glycerol solution as the solvent, another indication that the benzyl alcohol was not compatible with glycerol in Experiment B. Further, the effect of benzyl alcohol was minimal on the samples pretreated with soybean oil, but benzyl alcohol significantly improved dyeing without the soybean oil pretreatment, also indicating that benzyl alcohol was not compatible with soybean oil. Without a subsequent NTP treatment, soybean oil soaking resulted in slightly lower color strength, which seems contradictory to the results in Experiment B. However, Experiment C was carried out at a much lower temperature (T = 90 °C vs. T = 140 °C in experiment B with water or 50% aqueous glycerol solution (vs. 50% or 80% aqueous glycerol in Experiment B in the dye bath, and these factors may have limited the beneficial effects of the soybean oil in achieving dyeing without an NTP treatment. The longer NTP treatment time in Experiment C (90 s vs. 60 s in Experiment B also significantly improved dyeing even without the soaking in soybean oil pretreatment.

From the results of the previous Experiments A–C, it is clear that the use of glycerol in the dye solvent does not help with dyeing, and there is a potential for improving dyeing using a pretreatment consisting of soaking in soybean oil followed by NTP. Accordingly, Experiment D used a lower dyeing temperature (T = 90 °C), water as the dye solvent, a wider range of NTP treatment times (t = 30–150 s), eliminated the swelling agent benzyl alcohol as a factor and added surfactant as a new factor using TWEEN 80, because of its reported benefit in dyeing aramids [25,26]. Results of the 3-factor 2-level full factorial design are presented in Table 4.

Table 4. Experiment D: experimental design and the results [@].

Sample	Surfactant	Pretreatment	NTP Time (s)	K/S	K/S Std *
D1	None	None	30	0.81	0.006
D2	None	Soy oil	30	1.01	0.008
D3	None	None	150	0.97	0.008
D4	None	Soy oil	150	1.20	0.017
D5	TWEEN	None	30	0.61	0.010
D6	TWEEN	Soy oil	30	0.60	0.002
D7	TWEEN	None	150	0.79	0.003
D8	TWEEN	Soy oil	150	0.84	0.005

[@] Experiment carried out with soaking in soybean oil for time = 15 h, dye solvent = water, dye concentration = 0.1 wt.%, T = 90 °C and dyeing time = 1 h. * Mean and standard error of 3 replicates.

ANOVA for the factorial design indicated a significant model (F-test, $p < 0.0001$). All three of the factors and their two-way interactions were significant (t-test, $p < 0.0002$). Since all of the factors are involved in interactions, only the interaction plots are shown in Figure 5.

According to the half-normal probability plot, the surfactant TWEEN 80 had the highest impact on the dyeing color strength, followed by NTP treatment time, then soybean oil pretreatment in descending order. The use of TWEEN 80 resulted in weaker color strength (Figure 5). The surfactant, although helping to form a dye dispersion, might also hinder the diffusion of the dye onto the para-aramid fiber fabrics, even in the presence of soybean oil. The soybean oil pretreatment improved dyeing, which was further improved by longer subsequent NTP treatment times. Evidently, the longer NTP treatment time of 150 s, comparing to the treatment time of 30 s, resulted in higher color strength in the presence or the absence of TWEEN 80 surfactant.

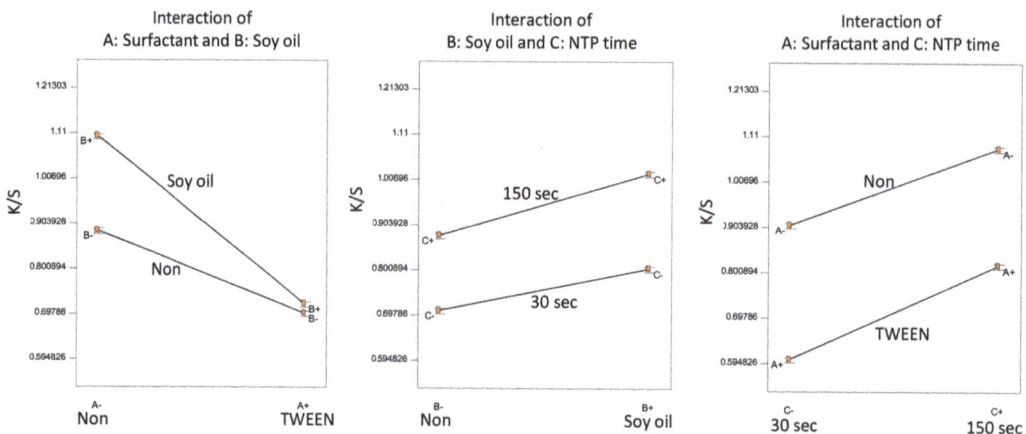

Figure 5. Statistical inference of Experiment D: significant factorial interactions.

It appeared that the NTP treatment time is crucial to the dyed color strength, especially for the soybean oil pretreated samples. Therefore, Experiment E was designed to optimize the NTP treatment time. In Experiment E, samples were soaked in soybean oil for 15 h, water was used as the dye solvent with a 0.1 wt.% dye concentration and with no other additives and the dyeing temperature was T = 90 °C for 1 h.

Results of the one-factor factorial design with NTP treatment time varied from 0–150 s are presented in Figure 6. The effects of NTP treatment at these experimental conditions show (Figure 6) that an NTP treatment time of 120 s provided the optimal result in dyeing the soybean oil pretreated samples in terms of color strength as represented by K/S values. NTP treatment times shorter than 120 s achieved dyeing, but to a lesser extent and NTP treatment times longer than 120 s decreased the K/S value, probably due to the degradation of the soybean oil network by the increased exposure to the NTP treatment.

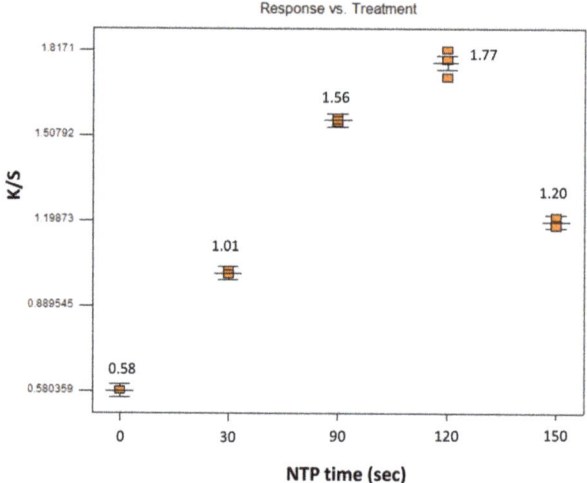

Figure 6. Effect of NTP time on K/S value of soybean oil-soaked para-aramids.

Based on the results of Experiments A–E, Experiment F was designed to optimize dyeing performance by developing a suitable recipe that included auxiliary additives (swelling agent and acetic acid) in the dye bath and a suitable dyeing process that in terms of the dyeing temperature (T = 70 °C or 100 °C) and the effects of a 120 s NTP treatment prior to versus post soaking in soybean oil. The addition of acetic acid into the dye bath brought down the pH to about 3, which should make the conditions favorable for cationic dyes [14]. Results of the 4-factor full factorial design are shown in Table 5.

Table 5. Experiment F: experimental design and the results [@].

Sample	Temperature	Acetic Acid	Swelling Agent	NTP	K/S	K/S Std *	ΔK/S [#]
F1	70 °C	Yes	Benzyl–OH	Post	1.08	0.016	−0.08
F2	70 °C	Yes	Benzyl–OH	Prior	0.91	0.022	−0.13
F3	70 °C	Yes	None	Post	0.99	0.005	−0.17
F4	70 °C	Yes	None	Prior	0.91	0.045	−0.27
F5	70 °C	No	Benzyl–OH	Post	1.18	0.008	−0.15
F6	70 °C	No	Benzyl–OH	Prior	1.12	0.025	−0.27
F7	70 °C	No	None	Post	0.96	0.023	−0.06
F8	70 °C	No	None	Prior	0.93	0.007	−0.31
F9	100 °C	Yes	Benzyl–OH	Post	1.55	0.012	−0.04
F10	100 °C	Yes	Benzyl–OH	Prior	1.15	0.016	−0.20
F11	100 °C	Yes	None	Post	1.30	0.007	−0.10
F12	100 °C	Yes	None	Prior	1.06	0.020	−0.17
F13	100 °C	No	Benzyl–OH	Post	1.45	0.010	−0.12
F14	100 °C	No	Benzyl–OH	Prior	1.21	0.068	−0.16
F15	100 °C	No	None	Post	2.28	0.064	−0.09
F16	100 °C	No	None	Prior	1.01	0.013	−0.17

[@] Experiment carried out with soaking in soybean oil for time = 15 h, dye solvent = water, dye concentration = 0.1 wt.%, dyeing time = 1 h, Additive concentrations = 1 wt.% Acetic acid and/or 2 wt.% Benzyl alcohol and the NTP treatment time = 120 s prior to or post soaking in soybean oil for 15 h. * Mean and standard error of 3 replicates. [#] The change in K/S value was calculated by subtracting the K/S after the 1st detergent wash from the K/S value measured after the 2nd detergent wash.

The half-normal probability plot indicated that impact of the factors is in the order of Dye temperature > NTP > Acetic acid. Even though the overall ANOVA model was significant ($p < 0.0001$), the coefficient for benzyl alcohol was not significant (t-test, $p = 0.63$). The only significant interaction term is Temperature × NTP (t-test, $p = 0.0003$), while the Temperature × Benzyl OH (Benzyl alcohol) interaction was marginally significant (t-test, $p = 0.085$). Figure 7 plots out the main effects of all the factors along with the two interactions.

Acetic acid was the only factor not involved in any significant interactions (Figure 7A). The addition of 1 wt.% acetic acid in the dye bath slightly decreased the dyeing color strength. All of the other three factors were involved in the two interactions, so their effects are discussed in the context of their interactions (Figure 7B).

The dyeing temperature and NTP after soaking in soybean oil significantly impacted the dyeing results, whereas the use of benzyl alcohol did not significantly influence the dyeing color strength. In Experiment F, the largest K/S value achieved was K/S = 2.28 (sample F15) in dyeing conditions with T = 100 °C in a water-based dye bath, with a pretreatment of soaking in soybean oil followed by a 120 s NTP treatment and no other additives of benzyl alcohol or acetic acid in the dye bath. It is noteworthy that T = 100 °C is the highest temperature that can be used for the water-based dye bath at atmospheric pressure without significant loss of water due to evaporation. In general, dyeing at T = 100 °C was better than dyeing at T = 70 °C. While NTP treatment (120 s) prior to soaking in soybean oil did not show a significant effect on dyeing, NTP treatment after soaking in soybean oil significantly improved the color strength, especially when the dyeing was carried out in the higher temperature regime of T = 100 °C. This clearly points out that the improved dyeing effect should be attributed to the chemical reactions between soybean oil and NTP.

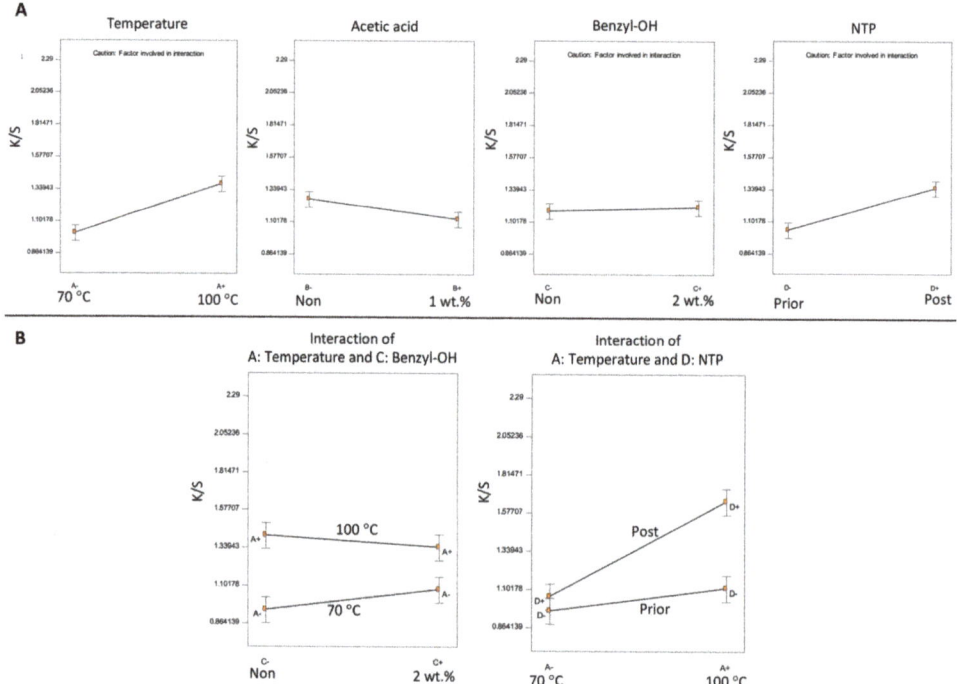

Figure 7. Analysis of Experiment F. (**A**) Main effects of all of the factors and (**B**) the significant interactions.

All samples were laundered with a detergent wash for a second time according to the same protocol described in Section 2.5, to determine colorfastness in terms of the change in K/S values (denoted as ΔK/S in Table 5). A negative sign of the ΔK/S value indicates a decrease in K/S value after the second wash with detergent. Generally, higher temperature and NTP treatment after soaking in soybean oil resulted in better dyeing strength and more durable dyeing (colorfastness) after laundering.

For the purposes of providing a visual demonstration comparing the dyeing strengths of different treatments, select samples were imaged using a flatbed scanner (Figure 8). Compared with the previous investigation of dyeing para-aramids with this method and using a disperse dye [27], it is demonstrated that the pretreatment of soybean oil followed by NTP treatment method is compatible with dyeing with both disperse dye and cationic dye. Further, this method enables dyeing to a significantly high color strength, although the disperse dye requires glycerol as a dispersant, but water as the solvent is better suited for the use of cationic dye. Taken together, these results demonstrate the potential for this method to replace hazardous chemicals currently used in dyeing practices with renewable materials and environmentally friendly ("green") technologies, to improve the dyeing of para-aramid textiles.

3.2. FTIR Analysis

As shown in Figure 9, the following characteristic peaks of soybean oil (line I) are well represented: 3009 cm^{-1} (=C–H stretching of aliphatic alkenes) in unsaturated fatty acids, 2922 cm^{-1} (–CH$_2$ asymmetric stretching), 2852 cm^{-1} (–CH$_3$ symmetric stretching), 1746 cm^{-1} (–C=O triglycerides carbonyl stretching), 1462 cm^{-1} (–CH$_2$ antisymmetric deformation) and 1160 cm^{-1} (C–O stretching in the esters) [30–33].

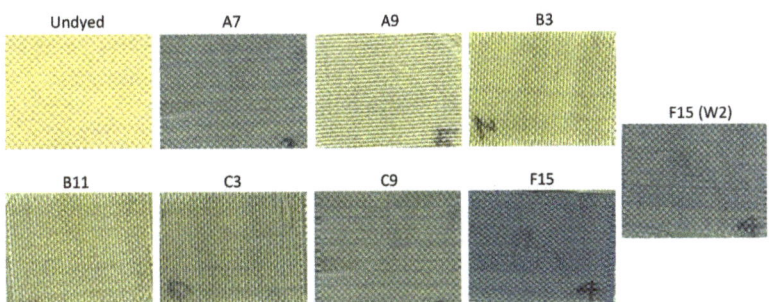

Figure 8. Scanned images comparing dyed samples with the undyed fabric (labels correspond to the sample numbers in Tables 1–5; F15 (W2) denotes sample F15 after second detergent wash).

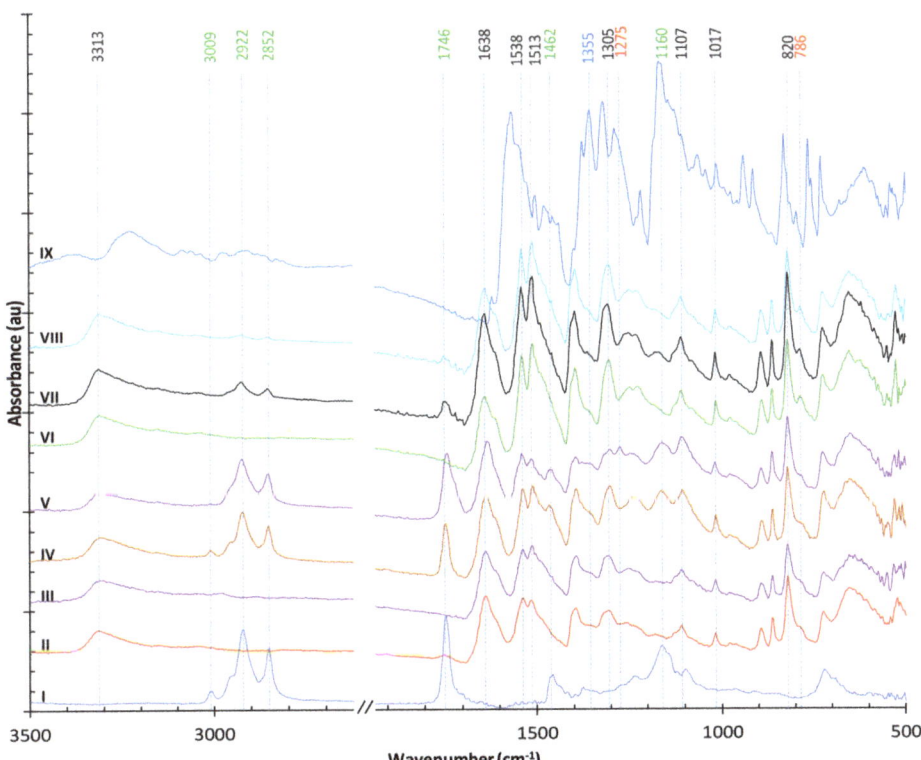

Figure 9. FTIR spectra of the para-aramid fabrics, materials and after certain treatments as follows: (**I**) Soybean oil; (**II**) Para-aramid fabric (untreated); (**III**) Para-aramid fabric treated with NTP; (**IV**) Para-aramid fabric after soaking in soybean oil; (**V**) Para-aramid fabric after soaking in soybean oil followed by a subsequent NTP treatment; (**VI**) Para-aramid fabric dyed with NTP as the only pretreatment; (**VII**) Para-aramid fabric dyed after soaking in soybean oil (without an NTP treatment); (**VIII**) Para-aramid fabric dyed after soaking in soybean oil and an NTP treatment; and (**IX**) Victoria Blue R cationic dye.

The spectrum of untreated para-aramids (line II) has main peaks at 3313 cm^{-1} (–N–H stretching), 1638 cm^{-1} (amide I C=O stretching) and 1538 cm^{-1} (–N–H deformation) [34]. The absorption bands at 1513 cm^{-1} (amide II), at 1017 cm^{-1} and at 820 cm^{-1} deriving from the C–H bonds on the para-aromatic rings and the absorption band at 1305 cm^{-1} deriving from the C–N bond stretch of amide III of the para-aramid fiber fabrics can be considered as internal standards for para-aramid [35]. NTP treatment of the para-aramid sample (line III) did not induce discernible changes in the observed FTIR peaks. A similar result was observed in Kašparová et al. [36], which also reported some changes in meta-aramids after NTP treatment and may reflect the different internal molecular structures between meta-aramids and para-aramids that make para-aramids more difficult to dye even with an NTP treatment.

The sample in line IV for the para-aramid fabric soaked in soybean oil shows peaks characteristic of soybean oil and of the para-aramids. Subsequent NTP treatment induces some important changes in the observed spectrum (line V). First, the unsaturated fatty acids peak at 3009 cm^{-1} decreases, but the other major characteristic peaks of soybean oil were still visible. This observation suggests that the double bonds in the unsaturated fatty acids were rapidly consumed by the NTP treatment, presumably by reaction with the plasma's reactive oxygen and nitrogen species (RONS). Second, there were changes in the 1380–1250 cm^{-1} region, including the formation of a small peak appearing at 1275 cm^{-1}, which is assigned to the C–C stretching vibration of the C–C(=O)–C groups in aliphatic ketone molecule [33]. These results suggest that the plasma's RONS attack the C=C double bonds, concomitant to the decrease in the peak at 3009 cm^{-1} (=C–H stretching of aliphatic alkenes). However, the ketone molecule is unstable. A likely explanation is that high-energy reactive oxygen species in the NTP (generated with ambient air as the feedgas) attack the electron rich double bonds in the unsaturated fatty acids and form epoxides, although the characteristic oxirane absorption peak typically observed at ~822 cm^{-1} would be obscured by the strong para-aramid absorption peak at 820 cm^{-1}. The –C=O triglycerides carbonyl stretching and C–O stretching in the esters at 1746 cm^{-1} and 1160 cm^{-1}, respectively, remain strong in line V, indicating that the NTP did not break the ester bonds in soybean oil.

Strong acids such as phosphoric acid and sulfuric acid have been used to pretreat para-aramids for the improvement of dyeing or interfacial bonding; a new absorption peak at ~3440 cm^{-1} (which is ascribed to the hydroxyl group O–H) or the shift and broadening of the 3313 cm^{-1} (–N–H stretching) band was reported, along with some decrease in mechanical strength. This distinct change was explained as a consequence of the increased number of –OH groups in the modified para-aramids due to hydrolysis [2,3]. However, this effect is not observed for the para-aramid sample pretreated with acetic acid in this study, and the acetic acid pretreatment did not significantly improve the color strength of dyeing.

FTIR spectra for dyed para-aramid fabric samples (Figure 9) are shown in line VI (with only NTP pretreatment), line VII (with soybean oil soaking but no NTP) and line VIII (with soybean oil soaking and NTP). The three spectra all have a new peak emerging at 786 cm^{-1}, which is typically the region of C–H bending for 1,3-disubstituted or 1,2,3-trisubstituted aromatic rings, possibly indicating a new bond forming on the para-aromatic rings. Otherwise, the spectrum of the sample dyed only with NTP pretreatment (line VI) is not much different from that of undyed para-aramids (line II or III). The para-aramid fabric sample in line VII (pretreated with soybean oil soaking and without subsequent NTP treatment, showed peaks characteristic of soybean oil that were significantly diminished. NTP treatment after oil soaking induced further decreases in the signal intensity of soybean oil (line VIII). In addition to reactions with NTP, there are two possible explanations for the diminished oil signal: (1) unfixed oil could be washed off after dyeing and subsequent washing with detergent; and (2) the triacylglycerol could be hydrolyzed at the dyeing conditions resulting in the destruction of ester bonds, as indicated by the diminished peaks at 1746 cm^{-1} and 1160 cm^{-1}.

The FTIR spectrum of the Victoria Blue R is also shown in Figure 9 (line IX). The para-aramid fabric samples showed a low uptake of dye, and most of the dye peaks were obscured by the characteristic peaks of the para-aramids and the oil. Accordingly, no significant dye peaks were detected in the dyed samples, except for a small shoulder at 1355 cm^{-1} in the spectra of the dyed samples (lines VII and VIII).

It is likely that the soybean oil well diffused throughout the tightly woven fabric and adsorbed onto the surfaces of yarns and fibers, and the NTP treatment induced the formation of a polymerized network in situ, enabling dyeing to a higher color strength. The unsaturated fatty acids in soybean oil play important roles in the dyeing process. Cross-linking/polymerization by the action of reactive plasma species are possible, because hydrogenation, nitration and epoxidation reactions have been observed along with polymerization under atmospheric NTP [37]. In a study on the tribological properties of air plasma polymerized soybean oil, Zhao et al. [33] concluded that the free radicals in the long-chain oil molecules, which were formed by the opening of the double bonds under the plasma conditions, could capture the reactive O and N species to produce and incorporate carbonyl, organic amine and nitrogen heterocyclic groups into the polymerized oil network. Moreover, depending on the duration of the NTP treatment in the dyeing procedure, the soybean oil might undergo processes such as complex oxidations, decomposition or fragmentation, to form reactive intermediate sites (e.g., ethers, furans, peroxides, carboxylic acids or 1,2,4-trioxolane [38,39]) on the para-aramid fiber fabrics that serve as catalysts or ligands to bind dye molecules.

4. Conclusions

Using a cationic dye and an environmentally friendly alternative dyeing method, we found that the best formulation for dyeing para-aramid textiles involved (i) pre-treating the para-aramid fiber fabrics with soybean oil followed by a brief NTP treatment, and (ii) using water at T = 100 °C as the dye solvent and omitting auxiliary chemical additives. Dyeing temperature and NTP treatment time were the most important factors, with the dyeing temperature of T = 100 °C resulting in higher color strength (larger K/S values), and the optimized NTP treatment time of 120 s for soybean oil pretreated para-aramids. These conditions achieved a K/S value up to 2.28, which is significantly larger than the K/S ~1 for untreated samples. For comparison, NTP alone or the use of the auxiliary additives of acetic acid and benzyl alcohol only slightly improved dyeing with the cationic dye. As a dye solvent, glycerol was inferior to water for dyeing with the cationic dye, and the addition of the surfactant TWEEN 80 in the dye bath negatively impacted the dyeing performance.

In the present study, FTIR analysis revealed that NTP induced chemical reactions in soybean oil on the surface of para-aramid fabrics were responsible for significantly improving the color strength. These chemical reactions likely involved high-energy, short-lived reactive plasma species that first attacked the double bonds in the unsaturated fatty acids. While the results of this study provide ample evidence that the pretreatment of soaking in soybean oil and ensuing NTP treatment improved dyeing color strength, determining the chemical pathways that led to better dyeing warrant further research, to control these processes with other dyestuffs. Future research along this line of sustainable dyeing method will aim at deriving more functionalities of soybean oil on the surface of para-aramids to further improve the dyeing strength and colorfastness.

Author Contributions: Conceptualization, C.J.D. and X.P.Y.; methodology, X.P.Y. and C.J.D.; validation, X.P.Y. and M.M.; formal analysis, C.J.D. and X.P.Y.; investigation, M.M. and X.P.Y.; resources, C.J.D. and X.P.Y.; data curation, M.M.; writing—original draft preparation, X.P.Y.; writing—review and editing, X.P.Y. and C.J.D.; visualization, M.M. and X.P.Y.; supervision, X.P.Y.; project administration, C.J.D. and X.P.Y.; funding acquisition, C.J.D. and X.P.Y. All authors have read and agreed to the published version of the manuscript.

Funding: This work was partially supported by the Department of Agriculture HATCH project No. TEN00521.

Institutional Review Board Statement: Not applicable.

Informed Consent Statement: Not applicable.

Data Availability Statement: The data presented in this study are available on request from the corresponding author.

Acknowledgments: We thank the support by the U.S. Department of Agriculture HATCH project No. TEN00521.

Conflicts of Interest: The authors declare no conflict of interest.

References

1. Roenbeck, M.R.; Sandoz-Rosado, E.J.; Cline, J.; Wu, V.; Moy, P.; Afshari, M.; Reichert, D.; Lustig, S.R.; Strawhecker, K.E. Probing the internal structures of Kevlar® fibers and their impacts on mechanical performance. *Polymer* **2017**, *128*, 200–210. [CrossRef]
2. Zhao, J. Effect of surface treatment on the structure and properties of para-aramid fibers by phosphoric acid. *Fibers Polym.* **2013**, *14*, 59–64. [CrossRef]
3. Xia, D.; Wang, L.J. Sulfuric Acid Treatment of Aramid Fiber for Improving the Cationic Dyeing Performance. *Adv. Mater. Res.* **2012**, *627*, 243–247. [CrossRef]
4. Vassiliadis, A.A.; Roulia, M.; Boussias, C.M. Disperse Dyeing Systems for p-aramid Fibers. In Proceedings of the 37th International Symposium on Novelties in Textiles, Ljubljana, Slovenia, 15–17 June 2006.
5. Dong, Y.; Jang, J. The enhanced cationic dyeability of ultraviolet/ozone-treated meta-aramid fabrics. *Color. Technol.* **2011**, *127*, 173–178. [CrossRef]
6. Jelil, R.A. A review of low-temperature plasma treatment of textile materials. *J. Mater. Sci.* **2015**, *50*, 5913–5943. [CrossRef]
7. Vu, N.; Michielsen, S. Near room temperature dyeing of m-aramid fabrics. *J. Appl. Polym. Sci.* **2019**, *136*, 48190. [CrossRef]
8. Han, S.Y.; Jaung, J.Y. Acid dyeing properties of meta-aramid fiber pretreated with PEO45-MeDMA derived from [2-(methacryloyloxy)ethyl] trimethylammonium chloride. *Fibers Polym.* **2009**, *10*, 461–465. [CrossRef]
9. Xi, M.; Li, Y.-L.; Shang, S.-Y.; Li, D.-H.; Yin, Y.-X.; Dai, X.-Y. Surface modification of aramid fiber by air DBD plasma at atmospheric pressure with continuous on-line processing. *Surf. Coat. Technol.* **2008**, *202*, 6029–6033. [CrossRef]
10. Sun, Y.; Liang, Q.; Chi, H.; Zhang, Y.; Shi, Y.; Fang, D.; Li, F. The application of gas plasma technologies in surface modification of aramid fiber. *Fibers Polym.* **2014**, *15*, 1–7. [CrossRef]
11. Su, M.; Gu, A.; Liang, G.; Yuan, L. The effect of oxygen-plasma treatment on Kevlar fibers and the properties of Kevlar fibers/bismaleimide composites. *Appl. Surf. Sci.* **2011**, *257*, 3158–3167. [CrossRef]
12. Abou Elmaaty, T.; Eltaweel, F.; Elsisi, H. Water-free Dyeing of Polyester and Nylon 6 Fabrics with Novel 2-Oxoacetohydrazonoyl Cyanide Derivatives under a Supercritical Carbon Dioxide Medium. *Fibers Polym.* **2018**, *19*, 887–893. [CrossRef]
13. Kaul, B.L. Mass pigmentation and solution dyeing of synthetic fibres. In *Synthetic Fibre Dyeing*; Hawkyard, C., Ed.; Society of Dyers and Colourists: Bradford, West Yorkshire, UK, 2004; p. 230.
14. Kim, E.-M.; Choi, J.-H. Dyeing Properties and Color. Fastness of 100% meta-Aramid Fiber. *Fibers Polym.* **2011**, *12*, 484–490. [CrossRef]
15. Ferrero, F.; Periolatto, M.; Rovero, G.; Giansetti, M. Alcohol-assisted dyeing processes: A chemical substitution study. *J. Clean. Prod.* **2011**, *19*, 1377–1384. [CrossRef]
16. Ferrero, F.; Periolatto, M. Glycerol in comparison with ethanol in alcohol-assisted dyeing. *J. Clean. Prod.* **2012**, *33*, 127–131. [CrossRef]
17. Pawar, S.S.; Maiti, S.; Biranje, S.; Kulkarni, K.; Adivarekar, R.V. A novel green approach for dyeing polyester using glycerine based eutectic solvent as a dyeing medium. *Heliyon* **2019**, *5*, e01606. [CrossRef] [PubMed]
18. Cheng, L.; Liu, L.; Ye, X.P. Acrolein Production from Crude Glycerol in Sub- and Super-Critical Water. *J. Am. Oil Chem. Soc.* **2013**, *90*, 601–610. [CrossRef]
19. Liu, L.; Ye, X.P. Simultaneous production of lactic acid and propylene glycol from glycerol using solid catalysts without external hydrogen. *Fuel Process. Technol.* **2015**, *137*, 55–65. [CrossRef]
20. Liu, L.; Ye, X.P.; Bozell, J.J. A Comparative Review of Petroleum-Based and Bio-Based Acrolein Production. *ChemSusChem* **2012**, *5*, 1162–1180. [CrossRef]
21. Zou, B.; Ren, S.; Ye, X.P. Glycerol Dehydration to Acrolein Catalyzed by ZSM-5 Zeolite in Supercritical Carbon Dioxide Medium. *ChemSusChem* **2016**, *9*, 3268–3271. [CrossRef]
22. Baliarsingh, S.; Jena, J.; Das, T.; Das, N.B. Role of cationic and anionic surfactants in textile dyeing with natural dyes extracted from waste plant materials and their potential antimicrobial properties. *Ind. Crops Prod.* **2013**, *50*, 618–624. [CrossRef]
23. Gur-Arieh, Z.; Ingamells, W. The Role of Benzyl Alcohol in the Solvent-assisted Dyeing of Acrylic Fibres. *J. Soc. Dye. Colour.* **2008**, *90*, 8–11. [CrossRef]
24. Shao, D.; Xu, C.; Wang, H.; Du, J. Enhancing the Dyeability of Polyimide Fibers with the Assistance of Swelling Agents. *Materials* **2019**, *12*, 347. [CrossRef] [PubMed]
25. Nechwatal, A.; Rossbach, V. The Carrier Effect in the m-Aramid Fiber/Cationic Dye/Benzyl Alcohol System. *Text. Res. J.* **1999**, *69*, 635–641. [CrossRef]
26. Zheng, H.-D.; Zhang, J.; Yan, J.; Zheng, L.-J. Investigations on the effect of carriers on meta-aramid fabric dyeing properties in supercritical carbon dioxide. *RSC Adv.* **2017**, *7*, 3470–3479. [CrossRef]

27. Morris, M.; Ye, X.P.; Doona, C.J. Soybean Oil and Nonthermal Plasma Pretreatment to Dye Para-Aramid Woven Fabrics with Disperse Dye Using a Glycerol-Based Dye Bath. *J. Am. Oil Chem. Soc.* **2021**, *98*, 463–473. [CrossRef]
28. Standard, ISO 105-C10:2006. Colour Fastness to Washing with Soap or Soap and Soda. In Textiles—Tests for Colour Fastness. In *Internaltional Standardisation Organisation*; Internaltional Standardisation Organisation: Geneva, Switzerland, 2006.
29. Etters, J.N.; Hurwitz, M.D. Opaque reflectance of translucent fabric. *Text. Chem. Color.* **1986**, *18*, 19–26.
30. Lumakso, F.; Rohman, A.M.H.; Riyanto, S.; Yusof, F. Detection and quantification of soybean and corn oils as adulterants in avocado oil using fourier transform mid infrared (FT-MIR) spectroscopy aided with multivariate calibration. *J. Teknol.* **2015**, *77*, 251–255. [CrossRef]
31. Man, Y.; Rohman, A. Analysis of Canola Oil in Virgin Coconut Oil Using FTIR Spectroscopy and Chemometrics. *J. Food Pharm. Sci.* **2013**, *1*, 5–9.
32. Sienkiewicz, A.M.; Czub, P. The unique activity of catalyst in the epoxidation of soybean oil and following reaction of epoxidized product with bisphenol A. *Ind. Crops Prod.* **2016**, *83*, 755–773. [CrossRef]
33. Zhao, X.; Yang, J.; Tao, D.; Xu, X. Synthesis and Tribological Properties of Air Plasma Polymerized Soybean Oil with N-Containing Structures. *J. Am. Oil Chem. Soc.* **2014**, *91*, 827–837. [CrossRef]
34. Haijuan, K.; Hui, S.; Jin, C.; Haiquan, D.; Xiaoma, D.; Mengmeng, Q.; Muhuo, Y.; Youfeng, Z. Improvement of adhesion of kevlar fabrics to epoxy by surface modification with acetic anhydride in supercritical carbon dioxide. *Polym. Compos.* **2019**, *40*, E920–E927. [CrossRef]
35. Mukherjee, M.; Kumar, S.; Bose, S.; Das, C.; Kharitonov, A. Study on the Mechanical, Rheological, and Morphological Properties of Short Kevlar™ Fiber/s-PS Composites. *Polym. Plast. Technol. Eng.* **2008**, *47*, 623–629. [CrossRef]
36. Kašparová, M.; Šašková, J.; Gregr, J.; Wiener, J. Using of DSCBD plasma for treatment of Kevlar and Nomex fibers. *Chem. Listy* **2008**, *102*, 1515–1518.
37. Ximena, V.Y. Characterization and Analysis of High Voltage Atmospheric Cold Plasma Treatment of Soilbean Oil. Ph.D. Thesis, Purdue University, West Lafayette, IN, USA, 2020.
38. Van Durme, J.; Nikiforov, A.; Vandamme, J.; Leys, C.; De Winne, A. Accelerated lipid oxidation using non-thermal plasma technology: Evaluation of volatile compounds. *Food Res. Int.* **2014**, *62*, 868–876. [CrossRef]
39. Zou, X.; Xu, M.; Pan, S.; Gan, L.; Zhang, S.; Chen, H.; Liu, D.; Lu, X.; Ostrikov, K.K. Plasma Activated Oil: Fast Production, Reactivity, Stability, and Wound Healing Application. *ACS Biomater. Sci. Eng.* **2019**, *5*, 1611–1622. [CrossRef] [PubMed]

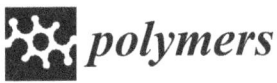

Article

A Negative Index Nonagonal CSRR Metamaterial-Based Compact Flexible Planar Monopole Antenna for Ultrawideband Applications Using Viscose-Wool Felt

Kabir Hossain [1,2], Thennarasan Sabapathy [1,2,*], Muzammil Jusoh [1,2], Mahmoud A. Abdelghany [3,4,*], Ping Jack Soh [1,5], Mohamed Nasrun Osman [1,2], Mohd Najib Mohd Yasin [1,2], Hasliza A. Rahim [1,2] and Samir Salem Al-Bawri [6]

[1] Advanced Communication Engineering (ACE), Centre of Excellence, Universiti Malaysia Perlis (UniMAP), Jalan Tiga, Pengkalan Jaya Business Centre, Kangar 01000, Malaysia; hossain.kabir42@gmail.com (K.H.); muzammil@unimap.edu.my (M.J.); pingjack.soh@oulu.fi (P.J.S.); nasrun@unimap.edu.my (M.N.O.); najibyasin@unimap.edu.my (M.N.M.Y.); haslizarahim@unimap.edu.my (H.A.R.)
[2] Faculty of Electronic Engineering Technology, Kampus Alam UniMAP Pauh Putra, Universiti Malaysia Perlis (UniMAP), Arau 02600, Malaysia
[3] Electrical Engineering Department, College of Engineering, Prince Sattam Bin Abdulaziz University, Wadi Addwasir 11991, Saudi Arabia
[4] Department of Electrical Engineering, Faculty of Engineering, Minia University, Minia 61519, Egypt
[5] Centre for Wireless Communications (CWC), University of Oulu, P.O. Box 4500, 90014 Oulu, Finland
[6] Space Science Centre, Climate Change Institute, Universiti Kebangsaan Malaysia, Bangi 43600, Malaysia; s.albawri@gmail.com
* Correspondence: thennarasan@unimap.edu.my (T.S.); abdelghany@mu.edu.eg (M.A.A.)

Abstract: In this paper, a compact textile ultrawideband (UWB) planar monopole antenna loaded with a metamaterial unit cell array (MTMUCA) structure with epsilon-negative (ENG) and near-zero refractive index (NZRI) properties is proposed. The proposed MTMUCA was constructed based on a combination of a rectangular- and a nonagonal-shaped unit cell. The size of the antenna was $0.825 \lambda_0 \times 0.75 \lambda_0 \times 0.075 \lambda_0$, whereas each MTMUCA was sized at $0.312 \lambda_0 \times 0.312 \lambda_0$, with respect to a free space wavelength of 7.5 GHz. The antenna was fabricated using viscose-wool felt due to its strong metal–polymer adhesion. A naturally available polymer, wool, and a human-made polymer, viscose, that was derived from regenerated cellulose fiber were used in the manufacturing of the adopted viscose-wool felt. The MTMUCA exhibits the characteristics of ENG, with a bandwidth (BW) of 11.68 GHz and an NZRI BW of 8.5 GHz. The MTMUCA was incorporated on the planar monopole to behave as a shunt LC resonator, and its working principles were described using an equivalent circuit. The results indicate a 10 dB impedance fractional bandwidth of 142% (from 2.55 to 15 GHz) in simulations, and 138.84% (from 2.63 to 14.57 GHz) in measurements obtained by the textile UWB antenna. A peak realized gain of 4.84 dBi and 4.4 dBi was achieved in simulations and measurements, respectively. A satisfactory agreement between simulations and experiments was achieved, indicating the potential of the proposed negative index metamaterial-based antenna for microwave applications.

Keywords: metamaterials; high-performance textiles; wearable antenna; textile antennas; polymer

1. Introduction

Flexible substrates including organic substances, such as polymers, paper, plastics, textiles, and fabrics, have become increasingly important to enable increased flexibility in wearable sensors/antennas [1]. Flexible antennas consist of a dielectric material (which works as the substrate) and a conductive material (which can be used as a radiating element and/or ground plane). Pure metals, metals mixed with fabrics, and conductive inks [2] are examples of materials that can be used as conductive materials. Meanwhile, polymers such as foam, paper textile fabrics, plastics, and soft printed circuit boards (PCBs) are all common dielectric polymer materials. Other than dielectric polymers, extensive research has been conducted on conductive polymers. They are explored for various applications such as solar energy harvesting [3], tissue engineering [4], supercapacitor design [5–7], gas sensors [8], and immunosensors [9]. In particular, flexible conductive polymers were also proposed by [10,11]. However, in antenna designs, more concerns are directed to the dielectric of the antenna since the antenna performance is mainly determined by the electrical characteristic and the mechanical flexibility of the dielectric substrate. Recent developments in manufacturing techniques for flexible polymer antennas are appealing due to the low permittivity with low losses [12]. In flexible antenna designs, dielectric polymer materials which are commonly used as substrates are classified either as natural polymers (e.g., rubber, silk, wool) or synthetic/human-made polymers (e.g., polystyrene, polyvinylchloride, nylon) [1]. Textile structural composites (which are considered as natural materials) are versatile in terms of their exceptional physical and mechanical properties which can be adopted in particular engineering applications to meet the desired requirements [13].

Ultrawideband (UWB) technology has triggered enormous research attention in wireless communications, especially after the allocation of the unlicensed frequency band (3.1 to 10.6 GHz) by the Federal Communications Commission (FCC) in 2002 [14]. A UWB antenna has the capability of providing high-speed data transmission with low-power spectral densities compared to conventional wireless communication systems within short distances. The application of UWB has been expanded into the wireless body area network (WBAN) domain based on the IEEE 802.15.6 WBAN standard [15,16]. Recent technological developments have resulted in compact and smart biomedical sensors/antennas for implementation on the human body. These antennas and sensors are most ideal for implementation in WBAN-type networks, as they are useful in sectors such as wearable computing, health monitoring, rescue systems, and patient tracking [17,18]. These applications require wireless devices to be placed close to the human body, which demands antennas and sensors to be developed using flexible materials. To prolong their usage near or on the body and, at the same time, ensure the safety and comfort of the user, they are best to be integrated onto clothing. Recently, conductive textiles have been introduced commercially, spurring the design of antennas for WBAN using textiles [16].

Sensors and smart devices have been the subject of extensive research over the past decade, with the goal of making them more easily integrated onto the human body [19]. Fabrics have been used as a natural and comfortable substrate for wearable electronic devices. Fabrics can now contain electrical functionality due to miniaturization of electronic components and innovative technologies [20]. There has been a lot of recent research on cloth fabrics, including sewn textiles, embroidered textiles, nonwoven textiles, knitted fabrics, woven fabrics, printed fabrics, braiding, laminated fabrics, spinning, and chemically treated fabrics [21]. Developing modern textile-based sensors has become a substantial undertaking in recent years, with numerous studies focusing on applications such as athletic training [22], emergency rescue and law enforcement [23], fitness monitoring [24], and other fields.

Metamaterials (MTMs) are artificial composite structures with exotic electromagnetic properties which can be used for potential groundbreaking applications (e.g., in antenna design, subwavelength imaging) [25]. Consequently, MTMs are suitable to be applied to improve WBAN antennas in terms of gain, radiation patterns, bandwidth (BW), and size compactness [26–28]. The characteristics of MTMs can be single negative (SNG) or double negative (DNG) based on the dielectric permittivity (ε) or magnetic permeability (μ). For SNG MTMs, either ε or μ can be negative, and for DNG MTMs, both ε and μ are negative. For SNG MTMs, if ε is negative, they are called epsilon-negative (ENG) MTMs, and if μ is negative, they are called mu-negative (MNG) MTMs [17,29]. Furthermore, the refractive index of a material depends on the ε and μ, which defines the extent of reflection and refraction [30]. However, the near-zero refractive index (NZRI) property can enhance the gain, as reported in [31]. Several metamaterial structures have been proposed in terms of complementary split-ring resonators (CSRR) [26,32], split-ring resonators (SRRs) [33], planar patterns, and capacitance-loaded strips (CLSs) [29]. Some other MTM structures such as electromagnetic bandgaps (EBGs) and artificial magnetic conductors (AMCs) were discussed in [27]. Such metamaterial-based UWB antennas have been reported in the literature with proven antenna performance enhancements [27,28]. In [34], MTMs were loaded into UWB wearable antennas for non-invasive skin cancer detection. Likewise, in [35], the proposed MTM UWB antenna was used for breast cancer detection. Despite the different designs, antennas for wearable applications should be compact, low cost, lightweight, and able to be integrated into circuits with ease [28]. When constructing metamaterials for metamaterial-enhanced devices, it is crucial to take into consideration the fabrication difficulty. Therefore, when developing textile-based metamaterials, extra care should be taken in each design phase [21,36].

This paper proposes a compact textile antenna incorporated with an MTM unit cell array (MTMUCA) structure, with an in-depth analysis. A polymer-based viscose-wool felt was adopted as the dielectric material of the antenna. The felt is a composite material that is developed from a naturally available polymer, wool, and a human-made polymer, viscose, that is derived from regenerated cellulose fiber. An equivalent circuit model was developed to present the working principles of the overall structure. Its structure was simulated and validated experimentally from 1 to 15 GHz. First, the transmission–reflection (RTR) method was used to extract the effective parameters of the MTMs in this work. The simulation results indicate that the MTM unit cell (MTMUC) and the MTMUCA are almost identical in performance, with an ENG BW of at least 11.53 GHz and an NZRI BW of 8.5 GHz. To the best of the authors' knowledge, the design of such textile-based ENG/ NZRI incorporated with a flexible MTM array antenna is yet to be reported in the literature. A comparison of the proposed antenna with similar designs in the literature is presented in Table 1.

Table 1. Comparison of the proposed design with relevant previous work in the literature.

Reference	Size (mm³) λ_0 = 40 mm	Operating Frequency Range (GHz)	Metamaterial Structure/Technique	Fractional Bandwidth (FBW) (%)	Antenna Peak Gain (dBi)	Remarks
[37]	52.5 × 52.5 × 20 (1.313 λ_0 × 1.313 λ_0 × 0.5 λ_0)	2.5–13.8 BW = 11.3	Flexible AMC metamaterial	138.65	9	Huge gap between the antenna and separate MTM layer made the overall antenna size bigger, complicated, and practically unusable.
[38]	43 × 40 × 2 (1.075 λ_0 × 1 λ_0 × 0.05 λ_0)	1.8–10 BW=8.2	Semicircular ring resonator in the patch	139	5.09	Comparatively low BW and large antenna size.
[39]	20 × 12 × 0.8 (0.5 λ_0 × 0.3 λ_0 × 0.02 λ_0)	2.4–10 BW = 7.6	Metamaterial	122.58	3.456	Antenna is compact in terms of size, but the BW and peak gain are not superior.
[17]	105 × 91 × 7.9 (2.625 λ_0 × 2.275 λ_0 × 0.1975 λ_0)	3.5–12.40 BW=8.9	Metasurface	114	9.1	Antenna size is large when including MTM. Increased design complexity with the use of multiple layered substrates.
[40]	50 × 43 × 14.95 (1.25 λ_0 × 1.075 λ_0 × 0.374 λ_0)	2.3–16 BW = 13.7	MNG metamaterial	149.73	8	The antenna has separate MNG MTM layer. Multilayer antenna design and huge gap between the antenna and MTM layer made the design complicated.
[34]	48 × 36 × 6 (1.2 λ_0 × 0.9 λ_0 × 0.15 λ_0)	8.2–13 BW=4.8	AMC metamaterial	45	7.04	Low BW and does not cover required FCC BW. Large antenna size incorporated on a separate AMC layer.
Proposed work	33 × 30 × 3 (0.825 λ_0 × 0.75 λ_0 × 0.075 λ_0)	2.55–15 BW= 12.45	SNG/NZRI metamaterial	142	4.84	Compact size and wide operational bandwidth and FBW.

2. Flexible Polymer-Based Textile Antenna Design with Metamaterial

In this work, the proposed MTM antenna was simulated and fabricated on textile materials. Shieldit Super™ with a thickness of 0.17 mm and a conductivity value of 1.18×10^5 S/m was used as the ground plane and the radiator. Meanwhile, a 3 mm viscose-wool felt substrate with a dielectric constant of 1.44 and loss tangent of 0.044 was employed. The choice of the viscose-wool felt was mainly due its strong Shieldit Super^{TM}–polymer adhesion. This type of flexible polymer is also easily available on the market; thus, no special treatment was required in developing the material in the lab, as required by other polymers such as polydimethylsiloxane (PDMS) [2]. The felt contained 30% viscose and 70% wool that formed a good composition of fibers with a density of 0.25 gm/CC. This property can help the Shieldit Super^{TM} easily iron out and attach to the polymer felt [36,41]. Computer Simulation Technology's (CST) Microwave Studio Suite (MWS) was used to model and simulate the MTM and MTM-integrated antenna over the frequencies of interest from 1 to 15 GHz. The analyses of these structures are reported in the following subsections.

2.1. Metamaterial Design

The proposed MTMUCA was designed based on a CSRR structure and is illustrated in Figure 1a. Its overall size was $12.5 \times 12.5 \times 3$ mm^3, and other dimensions are summarized in Table 2. A square loop and a nonagonal-shaped structure were combined to form the MTMUC structure. To characterize the metamaterial unit cell, the MTMUC structure was placed between two waveguide ports on the positive and negative z-axis and was excited with a transverse electromagnetic (TEM) wave, as depicted in Figure 2a. It was bounded by a perfect electric conductor (PEC) boundary at the $\pm x$-axis and a perfect magnetic conductor (PMC) boundary at the $\pm y$-axis. A frequency solver with a tetrahedral mesh scheme was utilized in simulations over the frequency of interest. In this metamaterial simulation, the unit cell was without a conductive layer at the bottom [42].

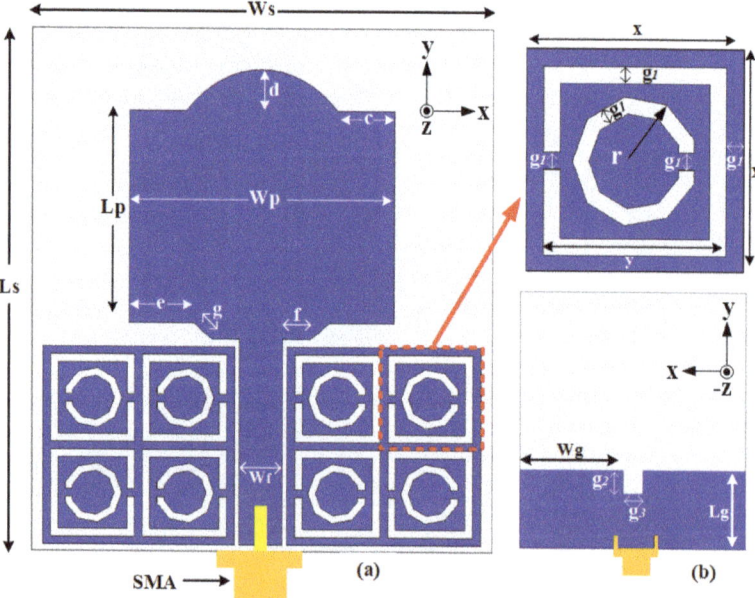

Figure 1. Schematic diagram of the proposed antenna and the MTMUCA structure (flexible polymer in gray and flexible conductive element in blue): (**a**) front view; (**b**) rear view.

Table 2. Parameter dimensions of the proposed antenna.

Para.	Value (mm)	Para.	Value (mm)	Para.	Value (mm)
Ls	33.00	c	3.68	x	6.50
Ws	30.00	d	2.60	y	5.50
Wp	17.20	e	4.20	r	1.75
Lp	13.40	f	1.81	g_2	3.00
Wg	13.70	g	1.56	g_3	2.60
Lg	10.00	g_1	0.5	Wf	2.80

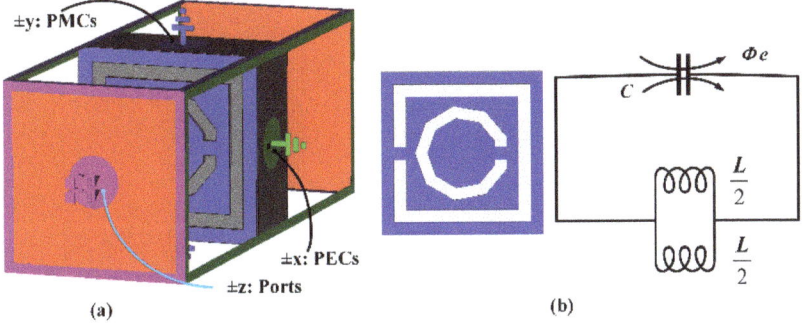

Figure 2. (a) 3D view of the MTMUCA simulation setup. (b) Topology of the MTMUC structure and its equivalent circuit model, where C = capacitance of MTMUCA, L/2 = each inductance (blue indicates the metallized areas).

The MTMUC structure and its equivalent circuit model are depicted in Figure 2b, where ohmic losses are unaccounted for [43]. The MTM modeled on a transverse plane acts as an *LC* resonator, which can be excited by the orthogonal electric field. Conversely, this structure behaves similarly to an electric dipole when excited by an axial electric field. The primary resonance can also be excited by the external magnetic field along the *y*-axis, as the CSRR can also exhibit a magnetic behavior [43,44]. The SNG properties can be tailored by appropriately modeling the CSRR gaps into the design. The surface current distribution of the proposed MTM was extracted for further study based on the setup exhibited in Figure 2a.

The simulated S-parameter of the MTMs is shown in Figure 3. In prior analyses, the material effective parameters (e.g., permittivity, refractive index) and MTM's stopband behavior were investigated. The transmission coefficient (S_{21}) of the MTMUC structure ranged from 1 to 4.65 GHz, and from 7.53 to 11.71 GHz, and the reflection coefficient (S_{11}) ranged from 6.34 to 6.68 GHz, and from 13.83 to 14.44 GHz. On the other hand, the MTMUCA showed an operational S_{21} from 1 to 4.45 GHz, and from 7.84 to 11.18 GHz, whereas the S_{11} was from 6.34 to 7.01 GHz, and from 13.65 to 14.26 GHz. Hence, the stopband behaviors clearly satisfied the $S_{21} \leq -10$ dB requirement.

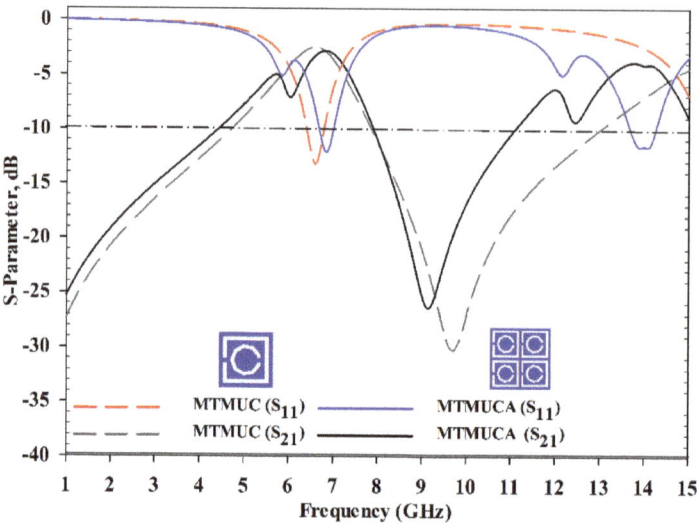

Figure 3. S-parameter of the MTMUC and MTMUCA structures.

The RTR [28,45] method was employed to extract the effective parameters from the normal incident scattering parameters using (1) to (7), starting with calculating S_{11} and S_{21} from Equations (1) and (2) as follows:

$$S_{11} = \left(\frac{R_{01}(1 - e^{i2nk_0d})}{1 - R_{01}^2 e^{i2nk_0d}}\right) \quad (1)$$

$$S_{21} = \left(\frac{(1 - R_{01}^2)e^{ink_0d}}{1 - R_{01}^2 e^{i2nk_0d}}\right) \quad (2)$$

where η is the refractive index, the wave vector in free space is denoted as k_0, the prototype/slab thickness is denoted as d, and $R_{01} = \frac{z-1}{z+1}$. Solving (1) and (2) results in (3) as follows:

$$z = \pm\sqrt{\frac{(1+S_{11})^2 - S_{21}^2}{(1-S_{11}^2) - S_{21}^2}} \quad (3)$$

$$e^{ink_0d} = X \pm i\sqrt{1 - X^2} \quad (4)$$

where $X = \frac{1}{2S_{21}(1-S_{11}^2+S_{21}^2)}$. As the material is deemed to be a passive medium, the impedance imaginary part should be greater than or equal to zero. Additionally, the real part of the refractive index (η) should be greater than or equal to zero. The η value of the material can be obtained from (5):

$$n = \frac{1}{k_0d}[\{imaginary(\ln e^{ink_0d}) + 2m\pi\} - i\{real(\ln e^{ink_0d})\}] \quad (5)$$

where m is an integer value or the branch index of the real part of η in other studies [46]. It can be noted that in this extraction method, $m = 0$ was considered [36]. The values of ε and μ can be determined from the following expression [46,47]:

$$\varepsilon = \frac{n}{z} \quad (6)$$

$$\mu = nz \quad (7)$$

Figure 4 illustrates the real part of the permittivity and refractive index. The MTMUC shows an ENG characteristic ($\varepsilon_r < 0$) from 1 to 6.45 GHz, from 7.48 to 13.52 GHz, and from 14.96 to 15 GHz, whereas its NZRI characteristic ($\eta < 0$) is featured from 1 to 6.16 GHz, and from 9.35 to 13.44 GHz. On the other hand, the MTMUCA displays an ENG characteristic from 1 to 6.75 GHz, from 7.70 to 13.36 GHz, and from 14.73 to 15 GHz, whereas the NZRI characteristic is featured from 1 to 5.74 GHz, 6.21 to 6.37 GHz, 9.48 to 12.08 GHz, and 12.49 to 13.19 GHz. Based on the S-parameters and the MTM characteristics, the proposed MTMUC and MTMUCA can be used for stopband applications and microwave applications [28], e.g., C-band and Ku-band.

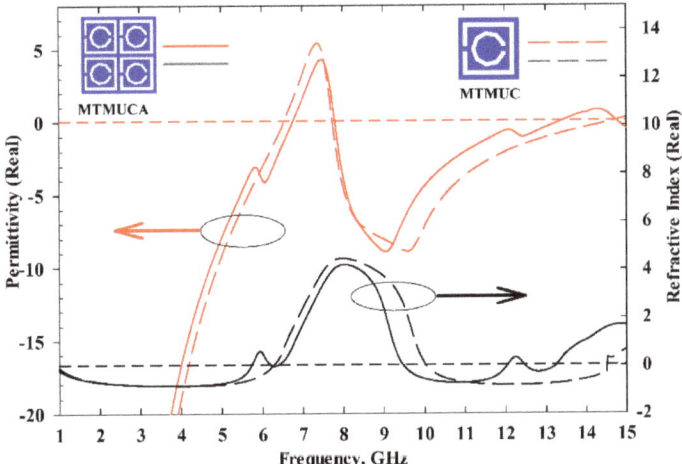

Figure 4. Permittivity and refractive index results of the MTMUC and MTMUCA structures.

2.2. Metamaterial Working Principle

To gain a further understanding of the MTM structure, a parametric study to analyze the effects of the nonagonal-shaped inner split conductor was performed, and its results are presented in Figure 5. In the absence of the inner conductor, a relatively narrower NZRI region can be observed for both the MTMUC and MTMUCA structures. Besides that, the structures' inner nonagonal-shaped split ring was further evaluated when rotated at 0°, 90°, 180°, and 270° angles. The results are summarized in Table 3. When positioned at 0° and 270° rotation angles, the structure shows identical results for both MTMUC and MTMUCA.

To provide further insight into the working principles and properties of the proposed metamaterial, the simulated current distribution was analyzed and discussed. Figure 6 visualizes the surface current distribution at 3 GHz, 6.5 GHz, 10 GHz, and 12 GHz for the different types of MTM structures. The surface current density and direction are indicated by the colors and arrows, respectively. The concentration of the surface current at 3 GHz is almost indistinguishable, as shown in Figure 6a–e for MTMUC and Figure 6f–j for MTMUCA. On the contrary, when the nonagonal-shaped inner split conductor is absent in Figure 6a,f, a lower surface current density at 6.5 GHz, 10 GHz, and 12 GHz is observed compared to the structures with an inner conductor. Stronger surface currents are also observed at the edges of the rectangles and the nonagonal-shaped slot. For the MTMUC structures (in Figure 6a–e), the concentration of the surface current in Figure 6c shows the strongest surface current distribution. Likewise, among the MTMUCA structures (in Figure 6f–j), Figure 6h indicates the strongest surface current distribution. It is observed that in both cases, the position of the inner nonagonal slot shown in Figure 6c,h facilitated the achievement of stronger surface currents compared to the rest of the structures.

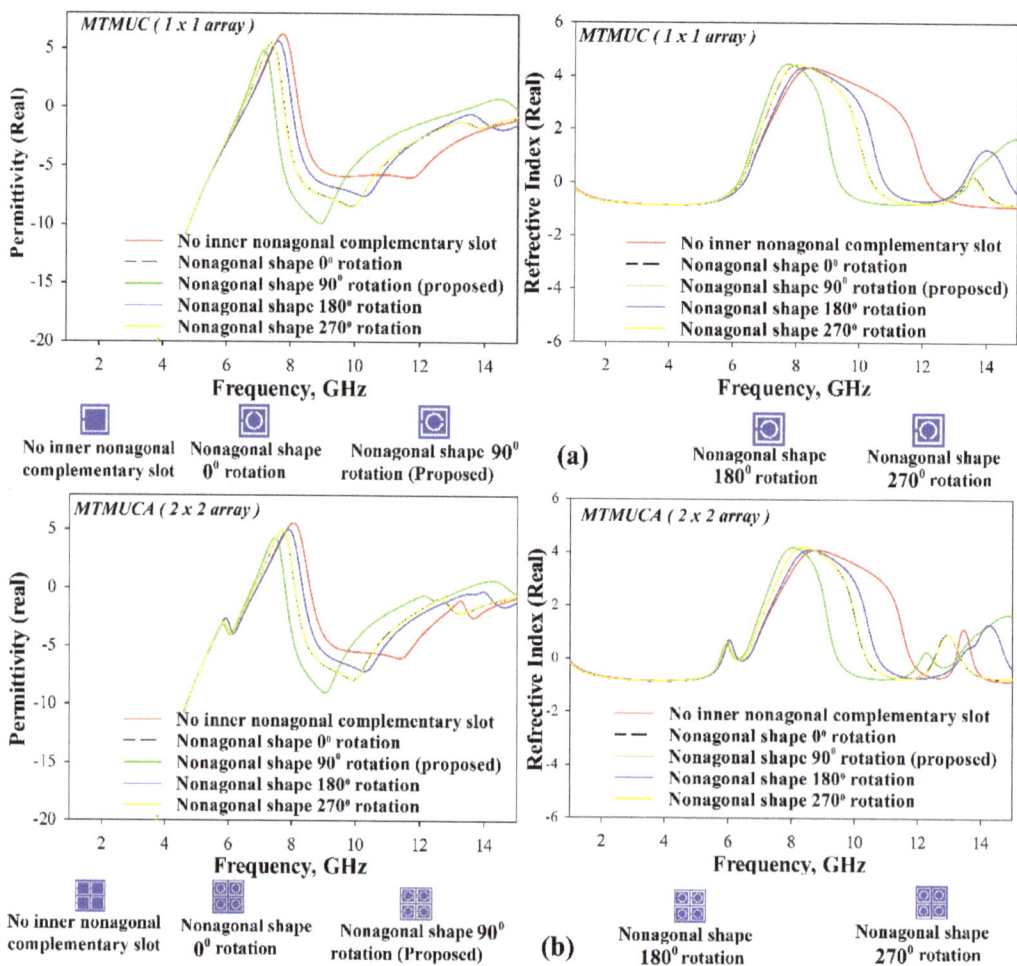

Figure 5. Analysis of relative permittivity and refractive index of (**a**) MTMUC and (**b**) MTMUCA.

Comparison of the structures with different rotation angles indicated that the 90° rotated structure showed the strongest surface current distribution for both MTMUC and MTMUCA. This structure was chosen over the 0° or 270° rotated structure despite the latter being slightly better in terms of ε and NZRI BW. Therefore, the chosen structure was implemented in the UWB antenna to be studied further in the following section.

Table 3. Negative permittivity and negative refractive index results of different types of MTM structures.

Array Structure	MTM Structure		
	Inner Nonagonal-Shaped Split Ring Rotation Angle (Degrees)	Negative Permittivity Band (GHz)	Negative Refractive Index Band (GHz)
1 × 1 (MTMUC)	–	1–6.58 8.2–15	1–6.33 12.17–15
	0	1–6.50 7.78–15	1–6.21 10.39–13.33 13.78–15
	90	1–6.45 7.48–13.52 14.96–15	1–6.16 9.35–13.44
	180	1–6.57 8.02–15	1–6.31 10.79–13.26 14.80–15.00
	270	1–6.50 7.78–15	1–6.21 10.39–13.33 13.78–15
2 × 2 (MTMUC$_A$)	–	1–6.88 8.57–15	1–5.79 6.28–6.56 11.79–13.19 13.74–15
	0	1–6.80 8.11–15	1–5.76 6.23–6.43 10.39–12.5 13.39–15
	90	1–6.75 7.70–13.36 14.73–15	1–5.71 6.21–6.37 9.48–12.08 12.49–13.19
	180	1–6.88 8.33–15	1–5.81 5.56–6.29 10.8–13.28 14.83–15
	270	1–6.8 8.11–15	1–5.76 6.23–6.43 10.39–12.5 13.39–15

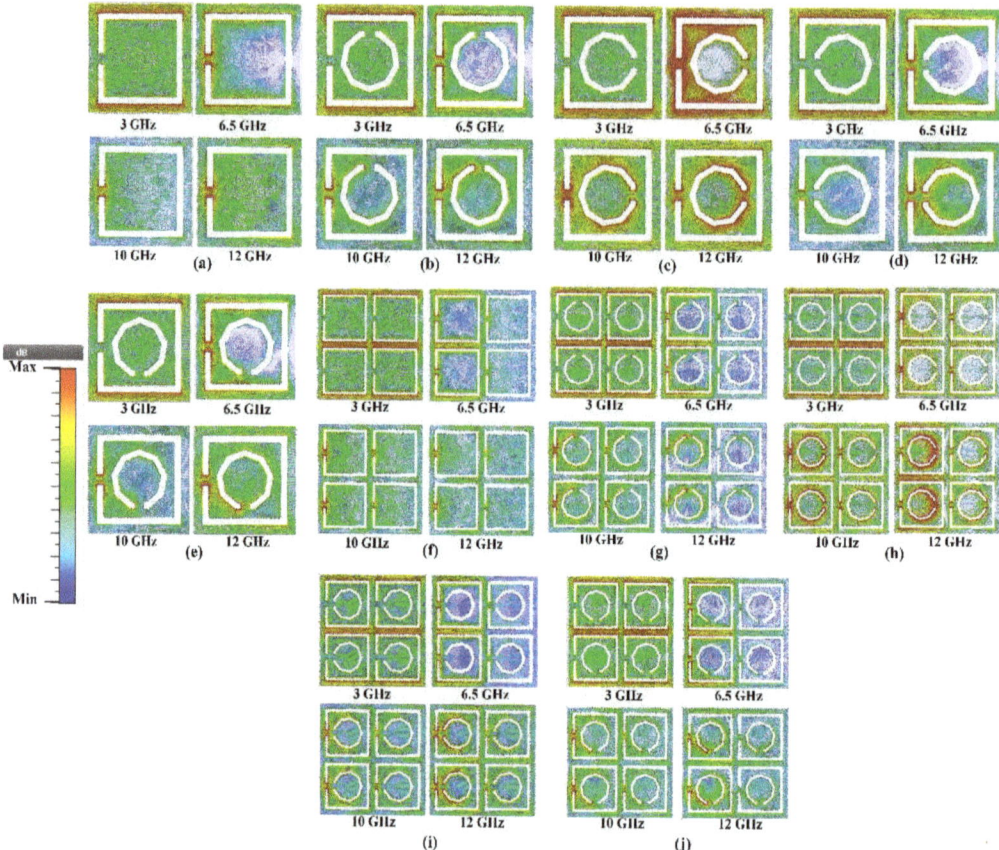

Figure 6. Surface current distribution for different MTM structures. MTMUC (1×1 array) (**a**) without the nonagonal-shaped inner split ring, and with (**b**) 0°, (**c**) 90°, (**d**) 180°, and (**e**) 270° rotation angles of the nonagonal-shaped inner split ring. MTMUCA (2 × 2 array) (**f**) without the nonagonal-shaped inner split ring, and with (**g**) 0°, (**h**) 90°, (**i**) 180°, and (**j**) 270° rotation angles of the nonagonal-shaped inner split ring.

2.3. Antenna Design Geometry and Configurations

As it has previously been mentioned, the structure of the proposed antenna integrated with the MTMUCA is depicted in Figure 1. The planar monopole antenna was designed with a combination of rectangular and half elliptical-shaped patches, whereas two MTMUCAs were located 0.4 mm from both sides of the planar feedline. A partial ground plane was implemented on the reverse side of this feedline, and a 50 Ω SMA connector was connected at the end of the feedline. The overall dimension of the antenna was 33 × 30 × 3 mm^3 (0.825λ_0 × 0.75λ_0 × 0.075λ_0, where λ_0 is the free space wavelength at 7.5 GHz, with the MTMUCA sized at 12.5 × 12.5 mm^2 (0.312λ_0 × 0.312λ_0)). All dimensions are summarized in Table 2.

Figure 7 depicts the surface current distribution at 3 GHz, 6.5 GHz, 10 GHz, and 12 GHz with and without the MTMUCA structure integrated into the antenna. It is evident that the MTMUCA improved the current intensity. The circuit model of the structure was modeled based on [44,48–50], where the conventional planar monopole antenna model is illustrated in Figure 8a. The planar monopole antenna's input impedance can be represented as an *RLC* circuit resonator near its resonance frequency, whereas the microstrip feedline can be expressed as a series inductor [44,48,50], resulting in the overall circuit model shown in Figure 8b. The CSRR-type MTMUC was modeled as a shunt *RLC* resonator tank (R_m, L_m, and C_m) [44], which was designed to work at the frequencies of interest. The resistor R_m indicates the dielectric and conductor losses, whereas the capacitance and inductance of the MTMCU are denoted as L_m and C_m, respectively. The short distance between the metallic ground plane and the MTMUCA was modeled as a capacitance, expressed as C_{MG} [51], whereas C_{MP} represents the capacitive coupling between the MTMUCA and microstrip feedline and/or patch resonator.

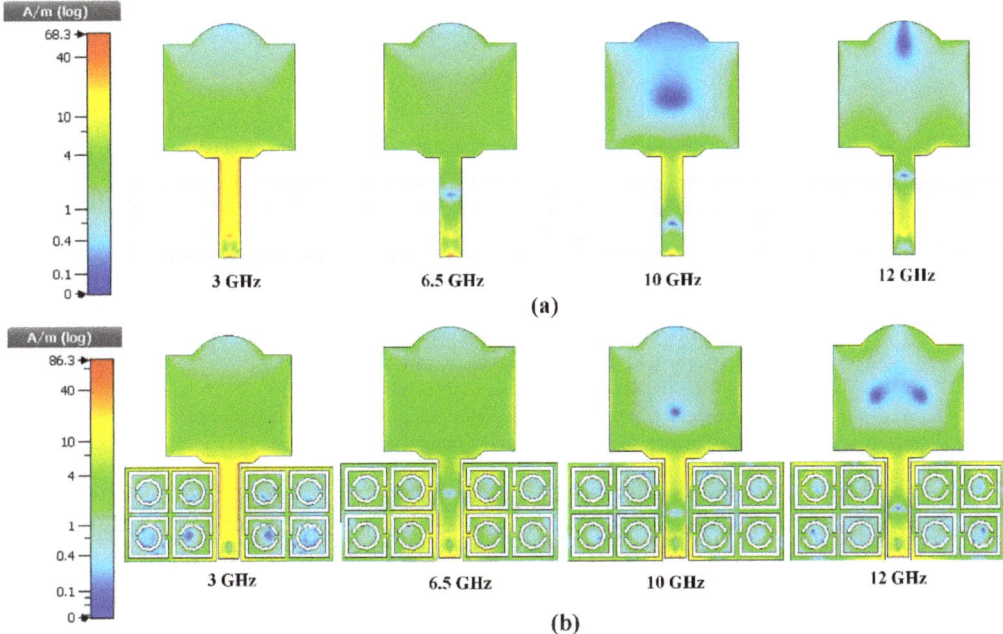

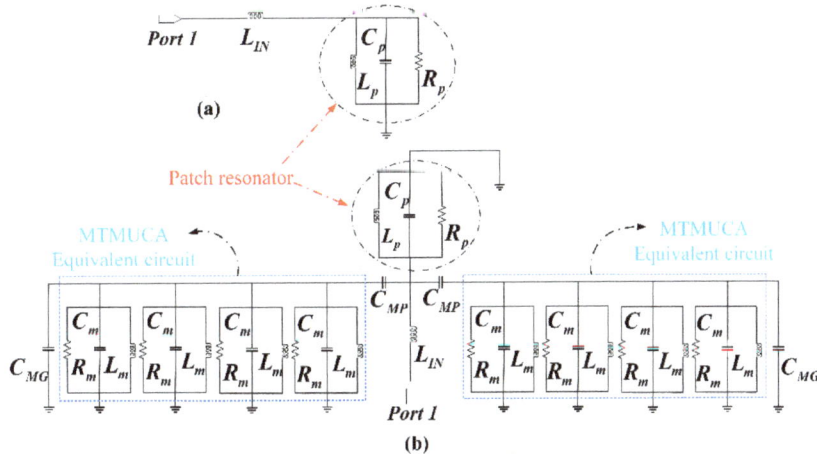

Figure 7. Surface current distribution. (**a**) Without the integration of the MTMUCA structure. (**b**) With the integration of the MTMUCA structure.

Figure 8. Equivalent circuit model. (**a**) Conventional planar monopole antenna with microstrip feedline antenna [48–50]. (**b**) Proposed antenna loaded with the MTMUCA demonstrated in Figure 1.

The microstrip feed line behaves as an inductance, denoted as L_{IN}. The CSRR acts as an electric dipole, and it mainly propagates along the xy plane within the substrate and radiates in the vicinity of the antenna.

To be more specific, the metamaterial's EM energy could be coupled to the planar monopole antenna through C_{MP}, whereas the radiation of the antenna was modeled using the radiation resistance R_p [44]. The same applies to the coupling between the feedline and the MTMUCA. The coupling effect can be seen in Figure 7b, where a significantly increased current concentration is observed. Therefore, the implementation of the MTMUCA enables the operation of the antenna at the frequencies of interest and will be demonstrated in the next section.

3. Results and Discussion

Figure 9 depicts the evolution process of the conventional antenna design in this study. In the first stage, as shown in Figure 9a, a full grounded patch with a combination of rectangular and half elliptical-shaped patches produces relatively narrow frequency bands, as shown in Figure 9b. In order to achieve a wider impedance bandwidth, the ground plane of the antenna was modified, as illustrated in the second stage of Figure 9b. From Figure 9b, it can be seen that the second stage has a wider impedance bandwidth compared to the first stage. Furthermore, in the third stage, a 0.4 mm gap was chosen to avoid a short circuit between the 50 Ω connector and the feedline of the planar monopole antenna. The truncation at the ground plane improved the impedance bandwidth slightly, where S_{11} resulted in being lower than −10 dB, showing over 2.33–2.6 GHz, and 8.52–12.3 GHz, which means FBW = 47% at the frequencies of interest. It can be noticed that the attained impedance bandwidth does not cover the complete UWB band allocated by the FCC. However, integration of the proposed MTMUCA with the conventional antenna can improve the overall performance. The related evidence can be found in the following results and discussion.

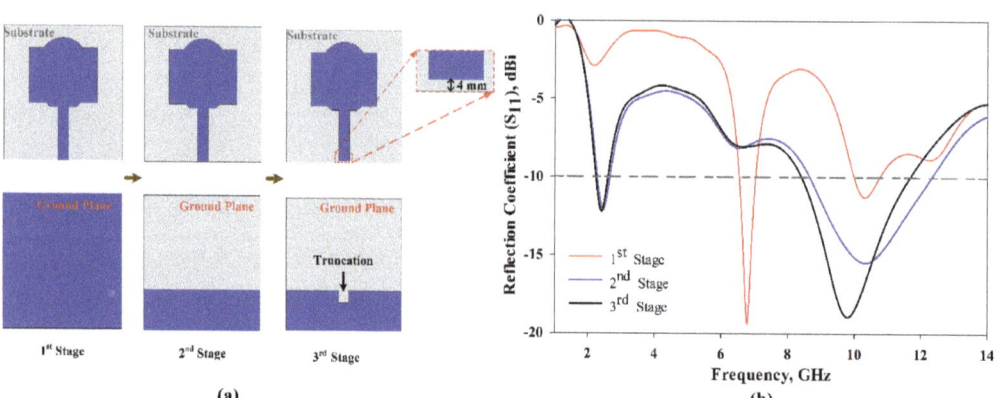

Figure 9. Evolution process of the conventional antenna in this study. (a) Evolution steps; (b) S_{11} results.

The fabricated prototype was used to validate this work, which is shown in Figure 10. The dielectric polymer material (viscose-wool felt) and conductive material (Shieldit Super™) of the prototype were dimensioned by a laser cutting machine as a part of the fabrication process. A great deal of study has already been carried out on a variety of materials that have characteristics that render them appropriate for use as a substrate for conductive materials for antennas, conductive threads [52], conductive polymers [53], and conductive textiles [54]. However, in this study, viscose-wool felt was adopted since it provides easier fabrication with sufficient flexibility and enables strong adhesion with the conductive textile Shieldit Super™. S_{11} measurement was performed using an Agilent E5071C Network Analyser (Agilent Technologies, Bayan Lepas, Penang, Malaysia) to ensure the simulated findings are accurate. The comparison of the proposed antenna between the simulated and measured S_{11} illustrated in Figure 11 indicates a good agreement. The simulation indicates a 10 dB impedance bandwidth from 2.55 to 15 GHz, which corresponds to an FBW of 142%. The measurement results indicate that this range is from 2.63 to 14.57 GHz, with an FBW of 138.84%. On the other hand, the antenna without the MTMUCA in Figure 9b (third iteration) shows a simulated FBW of 47%. Although the conventional antenna (considering the third iteration in Figure 9) does not work within the FCC region, by utilizing the unique characteristics of the metamaterial, the antenna performance could be enhanced. By utilizing the MTM on the conventional antenna, the antenna element's radiation efficiency, gain, and overall performance can be improved.

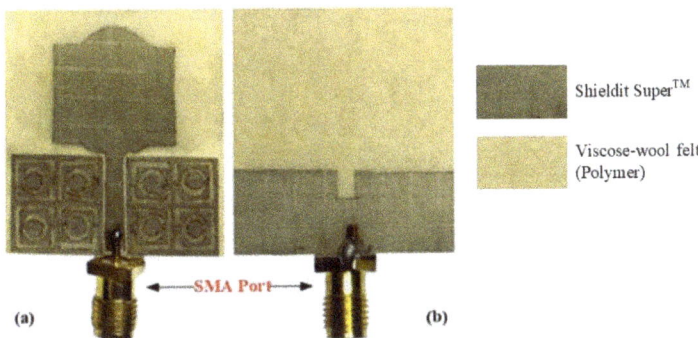

Figure 10. Prototype of the designed antenna: (**a**) front view, and (**b**) rear view.

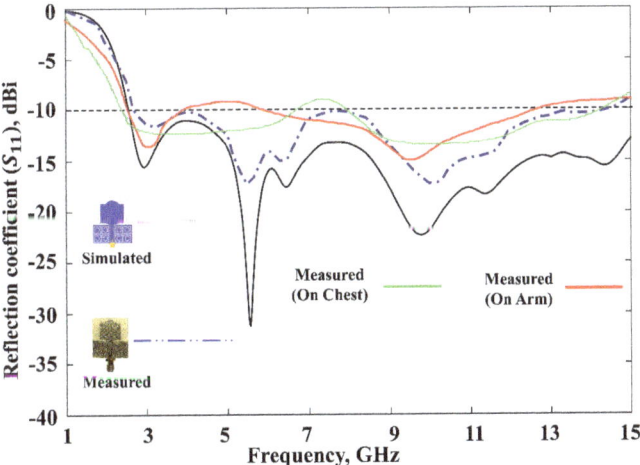

Figure 11. Reflection coefficients of the proposed antenna.

To rigorously analyze the performance of the proposed MTM-driven antenna prototype, body evaluation was further carried out with the help of a male volunteer (with a height of 1.72 m and weight of 84 kg), as shown in Figure 12. The prototype antenna was placed on two different places on the body (chest and arm). The related S_{11} results are presented in Figure 11, which indicate an excellent performance of the proposed antenna for on-body application. It can be noted that a 6 dB reflection coefficient is frequently used in the manufacturer's specification, as reported in [55,56]; hence, the obtained performance is sufficient for practical application.

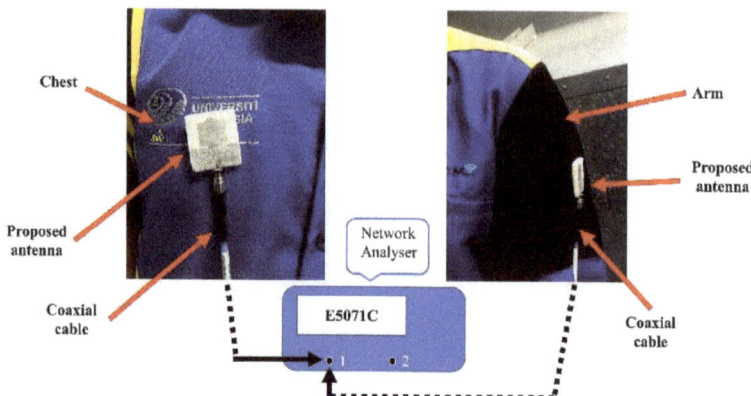

Figure 12. On-body measurement setup.

Figure 13 shows the antenna gain and total efficiency over the frequency. In simulations, an average realized gain of 3.3 dBi was achieved with the MTMUCA, whereas it was 2.7 dBi without the MTMUCA. A maximum gain of 4.83 dBi and 3.64 dBi could be obtained with the proposed antenna with and without the MTMUCA, respectively. On the other hand, the average measured gain of the proposed antenna is 3.04 dBi, and the maximum peak gain is 4.4 dBi. The maximum total efficiency achieved is 87% and 79.5% for the antenna with and without MTMUCA implementation, respectively, in simulations. The attainable average total efficiency of the proposed antenna is approximately 73%, whereas the antenna without the MTM indicated levels of around 69%. The measurements show a maximum total efficiency of 80%, and the average efficiency is 68.2%, for the proposed integrated antenna.

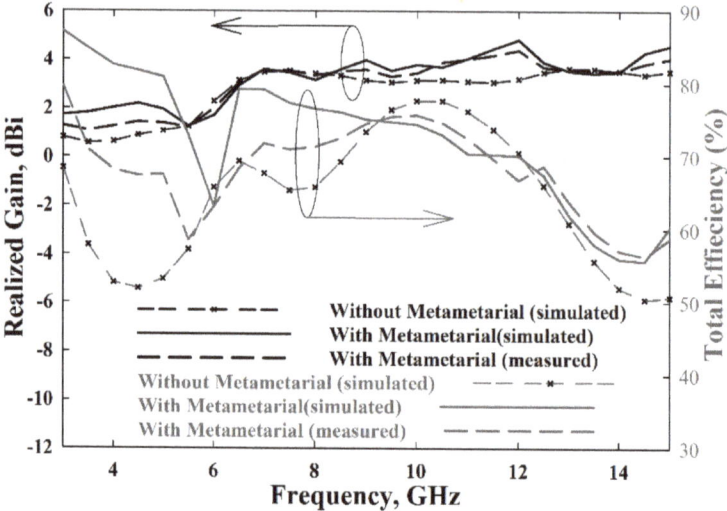

Figure 13. The proposed antenna's realized gain and radiation efficiency.

Figure 14 illustrates the radiation characteristics of the proposed antenna. The simulated and measured radiation patterns of the E-plane (yz-plane) and H-plane (xz-plane) were performed at four different frequencies, i.e., 3 GHz, 6.5 GHz, 10 GHz, and 12 GHz, indicating very good agreement. An omnidirectional radiation pattern can be observed at 3 GHz. Meanwhile, at 6.5 GHz, 10 GHz, and 12 GHz, an omnidirectional pattern can be seen in the H-plane, whereas the E-plane shows a bidirectional radiation pattern. Slight discrepancies between the simulations and measurements can be observed due to fabrication inaccuracies.

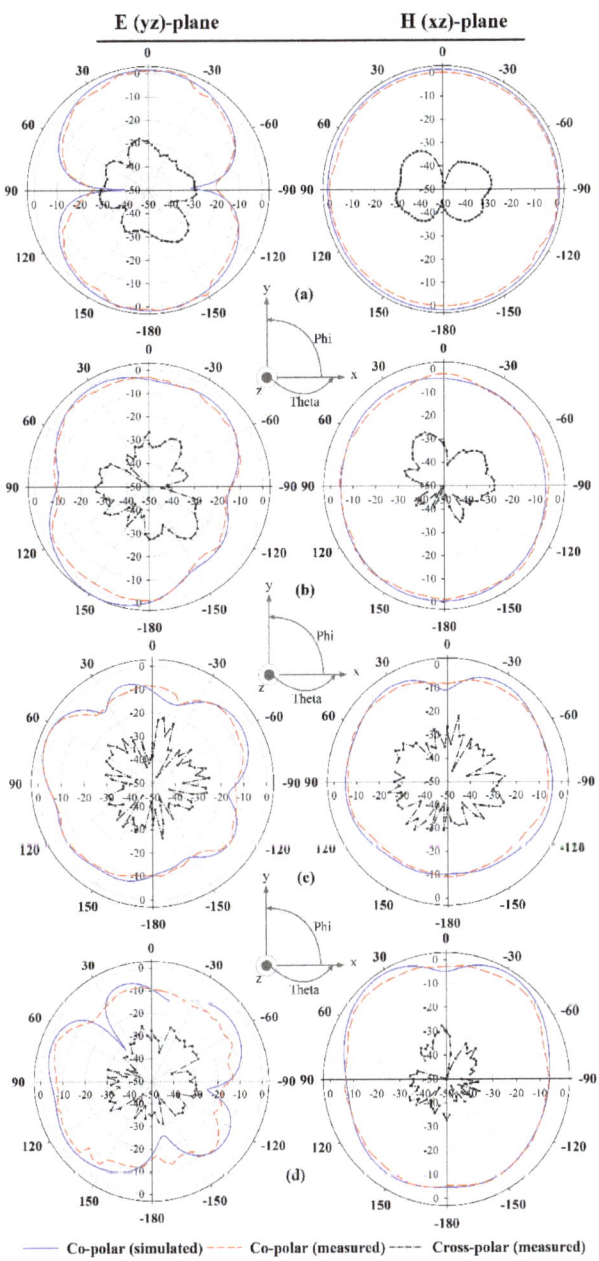

Figure 14. Radiation patterns of the proposed antenna at: (**a**) 3 GHz, (**b**) 6.5 GHz, (**c**) 10 GHz, and (**d**) 12 GHz.

4. Conclusions

A compact textile UWB planar monopole antenna integrated with an MTMUCA featuring ENG and NZRI properties was proposed and studied. The MTM-integrated antenna prototype was fabricated using flexible polymer viscose-wool felt as the substrate (since it provides easier fabrication with sufficient flexibility) and Shieldit Super^{TM} as the conductive element. The proposed MTMUC structure consists of a unique combination of square- and nonagonal-shaped CSRR-type MTMs. The proposed MTMUCA and MTMUC exhibited SNG properties in different frequency bands. The MTMUCA structure featured an 11.68 GHz ENG bandwidth and an 8.5 GHz NZRI bandwidth. This work also presented an equivalent circuit design to illustrate the principles of the overall structure. The implementation of the MTMUCA into the antenna design was proven experimentally to improve the antenna gain, efficiency, and bandwidth. Measurements showed an FBW of up to 138.84%, from 2.63 to 14.57 GHz, with a peak gain and total efficiency of 4.4 dBi and 80%, respectively. The proposed design can potentially be applied for wearable applications, where future research could be carried out to investigate the feasibility of on-body application, specifically for breast cancer detection.

Author Contributions: Conceptualization, K.H., M.J. and T.S.; design methodology, K.H., M.N.O. and P.J.S.; software, K.H. and H.A.R.; validation, K.H. and T.S.; formal analysis, K.H. and M.N.M.Y.; investigation, K.H.; design optimization, S.S.A.-B.; metamaterial characterization, K.H. and M.N.O.; numerical calculation and data interpretation, M.A.A. and H.A.R.; resources, M.N.O.; writing—original draft preparation, K.H., T.S. and M.J.; writing—review and editing, T.S., P.J.S. and S.S.A.-B.; visualization, K.H. and M.N.M.Y.; supervision, T.S. and M.J.; project administration, M.N.M.Y.; funding acquisition, M.A.A. All authors have read and agreed to the published version of the manuscript.

Funding: This work was supported by the Deanship of Scientific Research at Prince Sattam bin Abdulaziz University, Saudi Arabia.

Institutional Review Board Statement: Not applicable.

Informed Consent Statement: Not applicable.

Data Availability Statement: This study did not report any data.

Conflicts of Interest: The authors declare no conflict of interest.

References

1. Ali Khan, M.U.; Raad, R.; Tubbal, F.; Theoharis, P.I.; Liu, S.; Foroughi, J. Bending Analysis of Polymer-Based Flexible Antennas for Wearable, General IoT Applications: A Review. *Polymers* **2021**, *13*, 357. [CrossRef]
2. Kirtania, S.G.; Elger, A.W.; Hasan, M.R.; Wisniewska, A.; Sekhar, K.; Karacolak, T.; Sekhar, P.K. Flexible Antennas: A Review. *Micromachines* **2020**, *11*, 847. [CrossRef] [PubMed]
3. Farooq, S.; Tahir, A.A.; Krewer, U.; Shah, A.U.H.A.; Bilal, S. Efficient photocatalysis through conductive polymer coated FTO counter electrode in platinum free dye sensitized solar cells. *Electrochim. Acta* **2019**, *320*, 134544. [CrossRef]
4. Rahman, S.U.; Bilal, S.; ul Haq Ali Shah, A. Synthesis and Characterization of Polyaniline-Chitosan Patches with Enhanced Stability in Physiological Conditions. *Polymers* **2020**, *12*, 2870. [CrossRef]
5. Rahman, S.U.; Röse, P.; Surati, M.; Shah, A.U.H.A.; Krewer, U.; Bilal, S. 3D Polyaniline Nanofibers Anchored on Carbon Paper for High-Performance and Light-Weight Supercapacitors. *Polymers* **2020**, *12*, 2705. [CrossRef] [PubMed]
6. ur Rahman, S.; Röse, P.; ul Haq Ali Shah, A.; Krewer, U.; Bilal, S. An Amazingly Simple, Fast and Green Synthesis Route to Polyaniline Nanofibers for Efficient Energy Storage. *Polymers* **2020**, *12*, 2212. [CrossRef]
7. Ur Rahman, S.; Röse, P.; ul Haq Ali Shah, A.; Krewer, U.; Bilal, S.; Farooq, S. Exploring the Functional Properties of Sodium Phytate Doped Polyaniline Nanofibers Modified FTO Electrodes for High-Performance Binder Free Symmetric Supercapacitors. *Polymers* **2021**, *13*, 2329. [CrossRef]
8. Bibi, S.; Ullah, H.; Ahmad, S.M.; Ali Shah, A.-H.; Bilal, S.; Tahir, A.A.; Ayub, K. Molecular and Electronic Structure Elucidation of Polypyrrole Gas Sensors. *J. Phys. Chem. C* **2015**, *119*, 15994–16003. [CrossRef]
9. Zia, T.U.H.; Ali Shah, A.U.H. Understanding the adsorption of 1 NLB antibody on polyaniline nanotubes as a function of zeta potential and surface charge density for detection of hepatitis C core antigen: A label-free impedimetric immunosensor. *Colloids Surfaces A Physicochem. Eng. Asp.* **2021**, *626*, 127076. [CrossRef]
10. Jiang, X.; Xu, C.; Gao, T.; Bando, Y.; Golberg, D.; Dai, P.; Hu, M.; Ma, R.; Hu, Z.; Wang, X.-B. Flexible conductive polymer composite materials based on strutted graphene foam. *Compos. Commun.* **2021**, *25*, 100757. [CrossRef]

11. Zhou, K.; Dai, K.; Liu, C.; Shen, C. Flexible conductive polymer composites for smart wearable strain sensors. *SmartMat* **2020**, *1*, 1–5. [CrossRef]
12. Mohamadzade, B.; Hashmi, R.M.; Simorangkir, R.B.V.B.; Gharaei, R.; Ur Rehman, S.; Abbasi, Q.H. Recent Advances in Fabrication Methods for Flexible Antennas in Wearable Devices: State of the Art. *Sensors* **2019**, *19*, 2312. [CrossRef]
13. Jamshaid, H.; Mishra, R.; Zeeshan, M.; Zahid, B.; Basra, S.A.; Tichy, M.; Muller, M. Mechanical Performance of Knitted Hollow Composites from Recycled Cotton and Glass Fibers for Packaging Applications. *Polymers* **2021**, *13*, 2381. [CrossRef] [PubMed]
14. Commission, F.C. *FCC Report and Order for Part 15 Acceptance of Ultra Wideband (UWB) Systems from 3.1–10.6 GHz*; FCC: Washington, DC, USA, 2002; pp. 1–10.
15. Samal, P.B.; Soh, P.J.; Zakaria, Z. Compact Microstrip-Based Textile Antenna for 802.15.6 WBAN-UWB with Full Ground Plane. *Int. J. Antennas Propag.* **2019**, *2019*, 1–12. [CrossRef]
16. Samal, P.B.; Soh, P.J.; Xu, H.; Vandenbosch, G.A.E. Microstrip-based all-textile unidirectional UWB antenna with full ground plane. In Proceedings of the 8th European Conference on Antennas and Propagation, EuCAP 2014, Hague, The Netherlands, 6–11 April 2014; pp. 1408–1412. [CrossRef]
17. Yalduz, H.; Koç, B.; Kuzu, L.; Turkmen, M. An ultra-wide band low-SAR flexible metasurface-enabled antenna for WBAN applications. *Appl. Phys. A Mater. Sci. Process.* **2019**, *125*, 1–11. [CrossRef]
18. Gao, G.P.; Hu, B.; Wang, S.F.; Yang, C. Wearable Circular Ring Slot Antenna with EBG Structure for Wireless Body Area Network. *IEEE Antennas Wirel. Propag. Lett.* **2018**, *17*, 434–437. [CrossRef]
19. Tian, X.; Lee, P.M.; Tan, Y.J.; Wu, T.L.Y.; Yao, H.; Zhang, M.; Li, Z.; Ng, K.A.; Tee, B.C.K.; Ho, J.S. Wireless body sensor networks based on metamaterial textiles. *Nat. Electron.* **2019**, *2*, 243–251. [CrossRef]
20. Moradi, B.; Fernández-García, R.; Gil, I. E-Textile Embroidered Metamaterial Transmission Line for Signal Propagation Control. *Materials* **2018**, *11*, 955. [CrossRef]
21. Zhang, K.; Soh, P.J.; Yan, S. Meta-Wearable Antennas—A Review of Metamaterial Based Antennas in Wireless Body Area Networks. *Materials* **2020**, *14*, 149. [CrossRef]
22. Wu, J.F.; Qiu, C.; Wang, Y.; Zhao, R.; Cai, Z.P.; Zhao, X.G.; He, S.S.; Wang, F.; Wang, Q.; Li, J.Q. Human limb motion detection with novel flexible capacitive angle sensor based on conductive textile. *Electronics* **2018**, *7*, 192. [CrossRef]
23. Tartare, G.; Zeng, X.; Koehl, L. Development of a wearable system for monitoring the firefighter's physiological state. In Proceedings of the 2018 IEEE Industrial Cyber-Physical Systems ICPS 2018, St. Petersburg, Russia, 15–18 May 2018; pp. 561–566. [CrossRef]
24. Saenz-Cogollo, J.F.; Pau, M.; Fraboni, B.; Bonfiglio, A. Pressure mapping mat for tele-home care applications. *Sensors* **2016**, *16*, 365. [CrossRef]
25. Gric, T.; Hess, O. Metamaterial Cloaking. In *Phenomena of Optical Metamaterials*; Gric, T., Hess, O., Eds.; Elsevier: Amsterdam, The Netherlands, 2019; pp. 175–186. ISBN 9780128138960.
26. Ahdi Rezaeieh, S.; Antoniades, M.A.; Abbosh, A.M. Gain Enhancement of Wideband Metamaterial-Loaded Loop Antenna with Tightly Coupled Arc-Shaped Directors. *IEEE Trans. Antennas Propag.* **2017**, *65*, 2090–2095. [CrossRef]
27. Islam, M.M.T.; Islam, M.M.T.; Samsuzzaman, M.; Faruque, M.R.I. Compact metamaterial antenna for UWB applications. *Electron. Lett.* **2015**, *51*, 1222–1224. [CrossRef]
28. Al-Bawri, S.S.; Hwang Goh, H.; Islam, M.S.; Wong, H.Y.; Jamlos, M.F.; Narbudowiczm, A.; Jusoh, M.; Sabapathy, T.; Khan, R.; Islam, M.T. Compact Ultra-Wideband Monopole Antenna Loaded with Metamaterial. *Sensors* **2020**, *20*, 796. [CrossRef] [PubMed]
29. Khandelwal, M.K.; Arora, A.; Kumar, S.; Kim, K.W.; Choi, H.C. Dual band double negative (DNG) metamaterial with small frequency ratio. *J. Electromagn. Waves Appl.* **2018**, *32*, 2167–2181. [CrossRef]
30. Singh, H.; Sohi, B.S.; Gupta, A. A compact CRLH metamaterial with wide band negative index characteristics. *Bull. Mater. Sci.* **2019**, *42*, 1–11. [CrossRef]
31. Islam, S.S.; Iqbal Faruque, M.R.; Islam, M.T. Design and absorption analysis of a new multiband split-S-shaped metamaterial. *Sci. Eng. Compos. Mater.* **2017**, *24*, 139–148. [CrossRef]
32. Hossain, K.; Sabapathy, T.; Jusoh, M.; Soh, P.J.; Osman, M.N.; Al-Bawri, S.S. A Compact Wideband CSRR Near Zero Refractive Index and Epsilon Negative Metamaterial for Wearable Microwave Applications. *J. Phys. Conf. Ser.* **2021**, *1962*, 012019. [CrossRef]
33. Alemaryeen, A.; Noghanian, S. Crumpling effects and specific absorption rates of flexible AMC integrated antennas. *IET Microw. Antennas Propag.* **2018**, *12*, 627–635. [CrossRef]
34. Mersani, A.; Osman, L.; Ribero, J.M. Flexible UWB AMC antenna for early stage skin cancer identification. *Prog. Electromagn. Res. M* **2019**, *80*, 71–81. [CrossRef]
35. Islam, M.M.; Islam, M.T.; Faruque, M.R.I.; Samsuzzaman, M.; Misran, N.; Arshad, H. Microwave imaging sensor using compact metamaterial UWB antenna with a high correlation factor. *Materials* **2015**, *8*, 4631–4651. [CrossRef]
36. Hossain, K.; Sabapathy, T.; Jusoh, M.; Soh, P.J.; Jamaluddin, M.H.; Al-Bawri, S.S.; Osman, M.N.; Ahmad, R.B.; Rahim, H.A.; Mohd Yasin, M.N.; et al. Electrically Tunable Left-Handed Textile Metamaterial for Microwave Applications. *Materials* **2021**, *14*, 1274. [CrossRef]
37. Wang, F.; Arslan, T. A wearable ultra-wideband monopole antenna with flexible artificial magnetic conductor. In Proceedings of the 2016 Loughborough Antennas & Propagation Conference LAPC 2016, Loughborough, UK, 14–15 November 2016; pp. 3–7. [CrossRef]

38. Gogikar, S.; Chilukuri, S. A Compact Wearable Textile Antenna with Dual Band-Notched Characteristics for UWB Applications. In Proceedings of the 2019 9th IEEE-APS Topical Conference on Antennas and Propagation in Wireless Communications APWC 2019, Granada, Spain, 9–13 September 2019; 94, pp. 426–430. [CrossRef]
39. Elwi, T.A. Novel UWB printed metamaterial microstrip antenna based organic substrates for RF-energy harvesting applications. *AEU Int. J. Electron. Commun.* **2019**, *101*, 44–53. [CrossRef]
40. Negi, D.; Khanna, R.; Kaur, J. Design and performance analysis of a conformal CPW fed wideband antenna with Mu-Negative metamaterial for wearable applications. *Int. J. Microw. Wirel. Technol.* **2019**, *11*, 806–820. [CrossRef]
41. Sabapathy, T.; Soh, P.J.; Jusoh, M. A Three-Year Improvement Assessment of Project-Based Learning for an Antennas and Propagation Course [Education Corner]. *IEEE Antennas Propag. Mag.* **2020**, *62*, 76–84. [CrossRef]
42. Hossain, T.M.; Jamlos, M.A.F.; Jamlos, M.A.F.; Soh, P.J.; Islam, M.I.; Khan, R. Modified H-shaped DNG metamaterial for multiband microwave application. *Appl. Phys. A Mater. Sci. Process.* **2018**, *124*, 183. [CrossRef]
43. Baena, J.D.; Bonache, J.; Martín, F.; Sillero, R.M.; Falcone, F.; Lopetegi, T.; Laso, M.A.G.; García-García, J.; Gil, I.; Portillo, M.F.; et al. Equivalent-circuit models for split-ring resonators and complementary split-ring resonators coupled to planar transmission lines. *IEEE Trans. Microw. Theory Tech.* **2005**, *53*, 1451–1460. [CrossRef]
44. Dong, Y.; Toyao, H.; Itoh, T. Design and characterization of miniaturized patch antennas loaded with complementary split-ring resonators. *IEEE Trans. Antennas Propag.* **2012**, *60*, 772–785. [CrossRef]
45. Chen, X.; Grzegorczyk, T.M.; Wu, B.I.; Pacheco, J.; Kong, J.A. Robust method to retrieve the constitutive effective parameters of metamaterials. *Phys. Rev. E Stat. Physics Plasmas Fluids Relat. Interdiscip. Top.* **2004**, *70*, 7. [CrossRef]
46. Islam, S.S.; Faruque, M.R.I.; Islam, M.T. A new direct retrieval method of refractive index for the metamaterial. *Curr. Sci.* **2015**, *109*, 337–342.
47. Hossain, K.; Sabapathy, T.; Jusoh, M.; Soh, P.J.; Fazilah, A.F.M.; Halim, A.A.A.; Raghava, N.S.; Podilchak, S.K.; Schreurs, D.; Abbasi, Q.H. ENG and NZRI Characteristics of Decagonal-Shaped Metamaterial for Wearable Applications. In Proceedings of the 2020 International Conference on UK-China Emerging Technologies (UCET), Glasgow, UK, 20–21 August 2020; pp. 1–4.
48. Moradikordalivand, A.; Rahman, T.A.; Ebrahimi, S.; Hakimi, S. An equivalent circuit model for broadband modified rectangular microstrip-fed monopole antenna. *Wirel. Pers. Commun.* **2014**, *77*, 1363–1375. [CrossRef]
49. Wang, Y.; Li, J.Z.; Ran, L.X. An equivalent circuit modeling method for ultra-wideband antennas. *Prog. Electromagn. Res.* **2008**, *82*, 433–445. [CrossRef]
50. Balanis, C.A. *Antenna Theory: Analysis and Design*; John Wiley & Sons, 2016; ISBN 1118642066.
51. Navarro-Cía, M.; Beruete, M.; Agrafiotis, S.; Falcone, F.; Sorolla, M.; Maier, S.A. Broadband spoof plasmons and subwavelength electromagnetic energy confinement on ultrathin metafilms. *Opt. Express* **2009**, *17*, 18184. [CrossRef] [PubMed]
52. Huang, X.; Leng, T.; Zhang, X.; Chen, J.C.; Chang, K.H.; Geim, A.K.; Novoselov, K.S.; Hu, Z. Binder-free highly conductive graphene laminate for low cost printed radio frequency applications. *Appl. Phys. Lett.* **2015**, *106*, 203105. [CrossRef]
53. Sayem, A.S.M.; Simorangkir, R.B.V.B.; Esselle, K.P.; Hashmi, R.M. Development of Robust Transparent Conformal Antennas Based on Conductive Mesh-Polymer Composite for Unobtrusive Wearable Applications. *IEEE Trans. Antennas Propag.* **2019**, *67*, 7216–7224. [CrossRef]
54. Corchia, L.; Monti, G.; Tarricone, L. Durability of Wearable Antennas Based on Nonwoven Conductive Fabrics: Experimental Study on Resistance to Washing and Ironing. *Int. J. Antennas Propag.* **2018**, *2018*, 1–8. [CrossRef]
55. Guo, Z.; Tian, H.; Wang, X.; Luo, Q.; Ji, Y. Bandwidth enhancement of monopole uwb antenna with new slots and ebg structures. *IEEE Antennas Wirel. Propag. Lett.* **2013**, *12*, 1550–1553. [CrossRef]
56. Bialkowski, M.E.; Razali, A.R.; Boldaji, A. Design of an ultrawideband monopole antenna for portable radio transceiver. *IEEE Antennas Wirel. Propag. Lett.* **2010**, *9*, 554–557. [CrossRef]

Article

Fabrication of New Multifunctional Cotton/Lycra Composites Protective Textiles through Deposition of Nano Silica Coating

Tarek Abou Elmaaty [1,*], Hanan G. Elsisi [2], Ghada M. Elsayad [3], Hagar H. Elhadad [3], Khaled Sayed-Ahmed [4] and Maria Rosaria Plutino [5]

1. Department of Material Art, Galala University, Galala 43713, Egypt
2. Department of Textile Printing, Dyeing & Finishing, Faculty of Applied Art, Damietta University, Damietta 34512, Egypt; hanan.gamal58@yahoo.com
3. Department of Spinning, Weaving and Knitting, Faculty of Applied Art, Damietta University, Damietta 34512, Egypt; drghada3rm@yahoo.com (G.M.E.); hagarelhadad9@gmail.com (H.H.E.)
4. Department of Agricultural Chemistry, Faculty of Agriculture, Damietta University, Damietta 34512, Egypt; dr_khaled@yahoo.com
5. Stituto per lo Studio dei Materiali Nanostrutturati, Consiglio Nazionale delle Ricerche, Vill. S. Agata, 98166 Messina, Italy; plutino@pa.ismn.cnr.it
* Correspondence: tasaid@gu.edu.eg

Citation: Abou Elmaaty, T.; Elsisi, H.G.; Elsayad, G.M.; Elhadad, H.H.; Sayed-Ahmed, K.; Plutino, M.R. Fabrication of New Multifunctional Cotton/Lycra Composites Protective Textiles through Deposition of Nano Silica Coating. Polymers 2021, 13, 2888. https://doi.org/10.3390/polym13172888

Academic Editor: Giulio Malucelli

Received: 6 August 2021
Accepted: 25 August 2021
Published: 27 August 2021

Publisher's Note: MDPI stays neutral with regard to jurisdictional claims in published maps and institutional affiliations.

Copyright: © 2021 by the authors. Licensee MDPI, Basel, Switzerland. This article is an open access article distributed under the terms and conditions of the Creative Commons Attribution (CC BY) license (https://creativecommons.org/licenses/by/4.0/).

Abstract: This study aims to develop multifunctional pile cotton fabrics by implementing different compositions of lycra yarns with different densities of the cotton fabric under study. Highly dispersed silica nanoparticles (SiO_2 NPs) with small sizes—in the range of 10–40 nm—were successfully prepared and were analyzed using scanning electron microscopy (SEM). The particle size distribution of nano silica was determined via dynamic laser scattering (DLS) and measurements of its zeta potential. Cotton/lycra fabrics were treated using prepared SiO_2 NPs in presence of ethylenediaminetetraacetic acid (EDTA) as a crosslinking agent. Energy dispersive X-ray (EDX) analysis and scanning electron microscopy (SEM) were used to characterize the nano-treated fabrics and assure homogeneous dispersion of SiO_2 NPs on the cotton/lycra composites. Additionally, the nanoparticles were screened for their in vitro antibacterial activity against human pathogens such as Gram-positive *Staphylococcus aureus* and *Bacillus cereus* and Gram-negative *Escherichia coli* and *Pseudomonas aeruginosa* strains. The functional properties of the new composite pile cotton fabrics include excellent antibacterial, highly self-cleaning, and excellent UV protection factor (UPF) properties.

Keywords: cotton/lycra composites; silica nanoparticles; antibacterial activity; UV protection; self-cleaning

1. Introduction

Textiles are fundamental to a country's development and industrialization. As the demand for modern functional textiles grows, new materials and technology are being used. New multifunctional protective and smart textiles have been developed in response to growing technical breakthroughs, new standards, and a customer demand for textiles that are not only attractive but also practical. As a result, high-tech materials and well-considered fabric constructions will enhance wearer comfort while also providing unique features [1]. Textile products made from natural fibers such as cotton are good carriers of a variety of bacteria, which can cause health issues to the wearer [2]. Some of the most serious problems are "skin diseases," which can be developed by wearing contaminated clothes over a short time with the cross-transmission of bacteria found in air and on ground surfaces. Additionally, presence of these bacteria on textiles can result in undesirable damage to the fabric, such as fading, staining, a reduction in mechanical characteristics, and the material's deterioration [3].

Cotton is a widely used raw material in the manufacture of many fabrics, especially pile fabrics with many functional properties, such as water absorbency, humidity, durability when wet, and acceptable properties of skin friction, as well as the ability to withstand stress caused by washing and regular use [4]. Pile fabrics have a brush-like surface, which is created by tufts of warp or weft cut threads. A series of threads that protrude at right angles from a foundation or ground structure and form a pile or loop on the surface create the brush-like surface. It differs from other fabrics in regard to the surface texture (loops or cut ends) due to the extra warp and weft threads that appear at a certain height on the surface of the fabric according to the purpose of use, and the pile of these textiles comprises three threads [5]. Pile fabrics are distinguished from other textiles by their ability to illustrate and confirm the functional trends of the fabric's third dimension, represented by the thickness and height of the pile [6]. Obtaining the pile fabrics constitutes the greatest and most common part of the production of warp pile fabrics. These fabrics consist of two systems of warp threads for weaving warp pile fabrics (pile and ground warp), and one system of weft threads. There are methods of producing warp pile fabrics, including wire pile structures, that are used in upholstery, apparel wear, medical fabrics, ihram outfit clothes, curtains, etc.

However, the use of pure pile cotton fabrics does not achieve the required purpose of comfort and resistance to bacteria, which affects the performance of fabrics, attracts dust, and causes skin problems. Ahmed et al., 2020 [7] endeavored to solve the problems encountered by cotton fabrics using a facile fabrication of multifunctional cotton–modal–recycled aramid blended with protective textiles through the deposition of a three-dimensional tetrakis(hydroxymethyl)phosphonium chloride (THPC)-urea polymer coating. The results exhibited high antibacterial properties and superior water repellency.

On the other hand, synthetic fibers that can be used in the manufacture of pile fabrics such as lycra (Spandex) are seldom used alone. It is stretchy since it is a combination of nylon, cotton, or other fabrics [8]. Further, the use of lycra in woven fabrics provides a superior fit on the body, acting as a second skin, with good form retention and no distortion over the garment's lifetime. Badr 2017 [2] investigated the influence of fabric structure (Rib 4:1 Plain)/lycra combination on the antibacterial and mechanical properties. The findings revealed that the material type and fabric structure had an impact on the survival of *Escherichia coli* and *staphylococcus aureus* in socks. Lycra's further characteristics include less moisture absorption and resistance to both sunlight and industrial chemicals. The most common option among the textile is a combination of lycra yarn with different densities to impart a considerable level of stretch and recovery and to improve comfort due to its capacity to stretch, beyond that which can be achieved by cotton alone [9]. The elastomeric properties of a lycra and cotton blend are crucial in determining the elastic product's end-use [10,11]. Hence, we have focused on improving the functional performance of cotton fabrics by incorporating lycra yarns with different densities.

Performance apparel is exposed to a wide range of external conditions, including sunlight. Different factors promote the growth of melanoma and non-melanoma skin cancers, with UV radiation exposure being one of the most important. As a result, researchers have focused their efforts on altering fabric qualities as a layer to protect the skin from damaging radiation. Modifying the surface of fabrics to protect against UV radiation is crucial [12,13]. The relevance of self-cleaning arises when the fabric is continually exposed to dust and there is not enough time to wash them. One of the advantages of self-clean finishing is the elimination of traditional laundry procedures [14].

Nanotechnology has been successfully applied to various commercial products. It has also gained attention in the textile industry. Currently, it has been used in the processing and finishing of textiles to impart functional benefits [15]. Polymeric nanostructures, metal oxides, carbon nanotubes, clay nanoparticles (NPs), carbon black, graphite nanofibers, and other nanomaterials provide unprecedented textile performance, such as being hydrophilic and hydrophobic, antistatic, wrinkle-resistant, antimicrobial, antiodor, self-cleaning, and antiUV [16–21]. Inorganic NPs, especially TiO_2, ZnO, SiO_2, Cu_2O, CuO, Al_2O_3, and reduced graphene oxide NPs, are more commonly used than organic NPs due to their thermal and chemical durability at high temperatures, permanent stability under UV rays, and non-toxicity [22,23]. Nano silica has been proven to be a promising material due to its low density and good mechanical stability. Silica (SiO_2NPs) penetrate easily into a cotton fiber's interior and attach to the fiber structure tightly. As a result, the hydroxyl group of cellulose and SiOH in SiO_2NPs form a covalent bond [24–26]. This has recently been a significant research topic in both the scientific and industrial sectors. The addition of SiO_2 NPs to materials' surfaces increases their mechanical properties and durability, as well as influences their function, activity, and stability [27–29].

This study aims to examine the influence of both different fabric structure and implementation lycra yarns on the functional properties of the fabrics, as well as the treatment of the new fabric structure with highly dispersed SiO_2 NPs. During these processes, ethylenediaminetetraacetic acid (EDTA) was used as a crosslinking agent. It has numerous advantages: it is low cost, locally available, non-toxic, and provides the binding between SiO_2 NPs and cotton/lycra composites [23,30]. The SiO_2 treated cotton/lycra composites were also characterized, and their UV protective properties, antibacterial activities, and self-cleaning qualities were measured.

2. Materials and Methods
2.1. Fabric

The cotton fabric (100%), cotton/lycra (90.8/9.2%), and cotton/lycra (95.5/5.5%) employed have specifications given in Table 1. The weave structure used to prepare the samples was plain weave 1/2. The fabrics were prepared at (Cotton MISR Inc; Almahalla Al-Kubra, Egypt) using an electronic jacquard loom with the specification shown in Table 2. The samples were cleaned for 1 h at a boil with 2 g/L soap and 2 g/L sodium carbonate (Loba Chemie Pvt. Ltd., Boisar, India). The samples were washed in distilled water and air-dried.

Table 1. Specification of the prepared samples.

Fabric Specification	Material	Density/cm	Count (Ne)	Application Method
Warp specification	For ground warp: cotton is used for nine samples For Pile warp: cotton is used for nine samples	24 ends/cm (12 ground ends+12 pile ends)	24/2 Ne for both ground and pile warp ends	Wire pile
Weft Specification	Cotton (100%): for samples (A, B, and C) Cotton/Lycra (90.8/9.2%): for samples (D, E, and F) Cotton/Lycra (95.5/5.5%): for samples (G, H, and I)	(15, 19, 22) picks/cm	16/1 Ne	

Table 2. Specifications of the machine used for preparing samples.

Type of Loom	Picanol Nova 600
Manufacturing Country	Italy
Date of Manufacturing	1996
Reed count	12 dent/cm
Denting	2 Ends/Dent
Weft Insertion Device	Rapiers

2.2. Chemicals

Sodium metasilicate ($Na_2SiO_3 \cdot 9H_2O$, 95%, Merk KGaA, Darmstadt, Germany), cetyl trimethyl ammonium bromide (CTAB) ($C_{19}H_{42}BrN$, 98%, Loba Chemie Pvt. Ltd., Boisar, India), (HCl) (37%, Merk KGaA, Darmstadt, Germany), ethylene diamine tetra acetic acid (EDTA, Merk KGaA, Darmstadt, Germany), dihydrate, ($C_{10}H_{14}N_2Na_2O_8 \cdot 2H_2O$, 99.5%, J.T. Baker, Phillipsburg, NJ, USA), silver nitrate ($AgNO_3$, Merk KGaA, Darmstadt, Germany), and sodium chloride (NaCl, Merk KGaA, Darmstadt, Germany) were used in pure forms.

2.3. Preparation of Silica Nanoparticles (SiO$_2$ NPs)

SiO$_2$ NPs were prepared by adding 6 g sodium metasilicate to (394 mL) distilled water in a flask. The temperature was adjusted to 55 °C under constant stirring at a 300 rpm speed. CTAB (4 g) was added with stirring until complete dissolution for 15 min. The diluted HCl (200 mL) was added dropwise to the solution in two steps. In the first step, HCl was added dropwise to the mixture until the pH reached 9–9.5. With no further addition of HCl, the solution was agitated for another 10 min. More HCl was added until the pH reached 3–3.5 in the second step. Then, 10 mL of 10% NaCl aqueous solution was added to the reaction mixture at the acquired pH and agitated for an additional 20 min. The precipitated wet-gel silica was aged at 50 °C for 24 h, and the produced wet-gel silica was centrifuged and rinsed with distilled water until a negative response for chloride ions was observed, as measured with a 0.1 M AgNO$_3$ solution. The gel was dried at 40 °C via a microwave heating process. The total drying time was 70 min for each sample, applied seven times in an intermittent interval of 10 min. The dried gel was calcined at 650 °C for 3 h in a muffle furnace to obtain the white powder of nano SiO$_2$ [31,32]. The obtained micrographs showed that the SiO$_2$ NPs size was in the range of 10–40 nm, whereas their shape was spherical (Figure 1). Moreover, most SiO$_2$ NPs had a uniform size, and no aggregation was observed in scanning electron microscopy (SEM) analysis (JEOL JSM-6510LB with field emission gun, Tokyo, Japan). On the other hand, IR spectra were obtained for KBr pellets on a JASCO 410 spectrometer (Jasco, Tokyo, Japan) with only selected absorptions recorded in the range of 2000–200 cm^{-1}. Figure 2 shows that the peaks at 1020–1110 and 807 cm^{-1} are attributed to Si-O-Si a symmetric and symmetric stretching style, respectively [33].

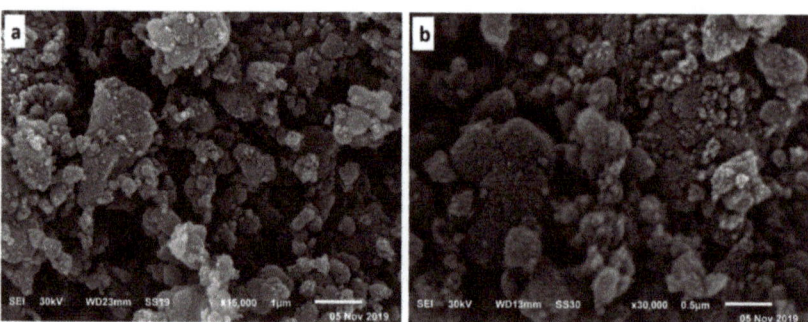

Figure 1. SEM micrographs of the amorphous SiO$_2$ NPs at different bar scales (a) 1 µm, and (b) 0.5 µm.

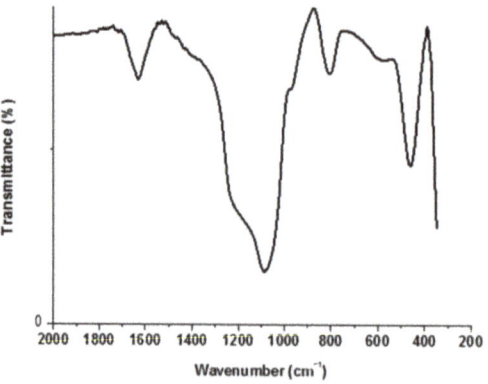

Figure 2. FTIR spectra of the amorphous silica prepared with CTAB.

Dynamic Light Scattering (DLS) and Zeta Potential Analysis

Zeta potential and dynamic light scattering (DLS) were measured using a Malvern Zetasize Nano-zs90. As shown in Figure 3, DLS analysis confirmed the preparation of SiO_2 particles in nano scale at approximately 90 nm. The obtained size from the DLS analysis is more than that obtained from SEM micrographs, which is caused by the formation of hydrogen bonds between SiO_2 NPs because of water addition [34]. Furthermore, the zeta potential value of the colloidal SiO_2 NPs solution was −19.4 mV with a single peak.

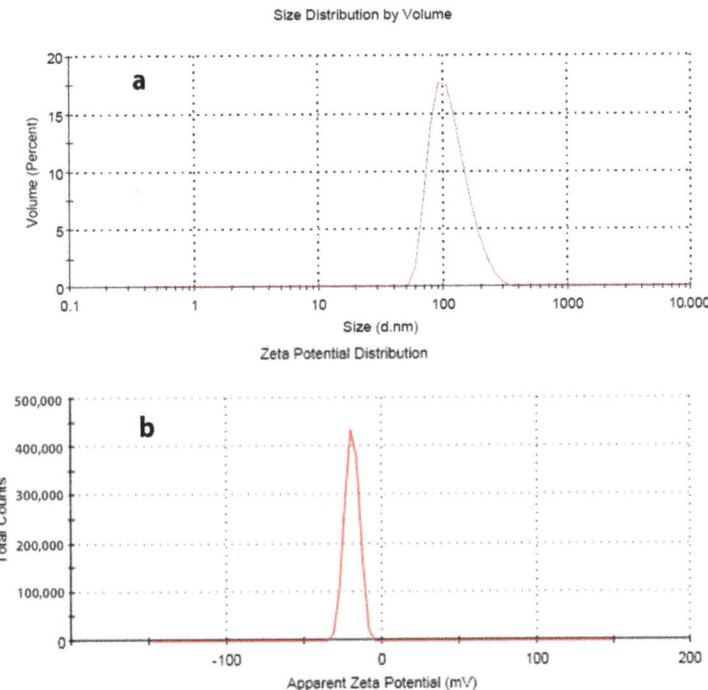

Figure 3. Dynamic light scattering (DLS) (**a**) and zeta potential (**b**) of the obtained SiO2 NPs.

2.4. Treatment of Fabric with SiO_2 NPs

The pad-dry-cure procedure was used to incorporate SiO_2 NPs into the cotton/lycra composites. EDTA (1 g, as a crosslinking agent) was added to 19 mL distilled water to prepare a nano silica solution. Afterwards, the mixture was agitated with a magnetic stirrer. Drops of silica amounting to 0.2 g were added to the liquid and stirred constantly until complete dissolution. The solution was sonicated for 30 min at room temperature. The fabric was soaked in the solution for 60 min followed by drying in an oven at 100 °C until completely dried, then cured at 120 °C for 60 min.

2.5. Characterization and Functional Properties of Fabrics Treated with Metal NPs

2.5.1. Scanning Electron Microscopy (SEM) Analysis

The surface morphologies of blank cotton/lycra composites and SiO_2 NP/cotton/lycra composite fabrics were characterized via scanning electron microscope (JEOL JSM-6510LB with field emission gun, Tokyo, Japan).

The deposition of SiO_2 NPs into cotton/lycra composites was confirmed using a surface energy dispersive X-ray (EDX) analysis unit (EDAX AMETEK analyzer, Mahwah, NJ, USA) attached to an SEM device. On the other hand, the elemental mapping analysis also reflects the presence of SiO_2 NPs on the surface of the treated cotton/lycra composites.

2.5.2. Antibacterial Activity

The antibacterial activity of the blank and SiO$_2$NP/cotton/lycra composite fabrics against Gram-positive bacteria *Staphylococcus aureus* and *Bacillus cereus* and Gram-negative bacteria *Escherichia coli* and *Pseudomonas aeruginosa* was assessed qualitatively using the AATCC Test Method (147-1988) expressed as the zone of growth (mm) [18,35].

2.5.3. UV Protection Properties

The blank and SiO$_2$NP/cotton/lycra composites were examined for UV protection functionality, expressed as UV protection factor (UPF), according to the Australian/New Zealand standard (AS/NZS 4399: 1996), and rated as follows: excellent protection (UPF > 40), very good protection (UPF: 25–39), and good protection (UPF: 15–24) [18,19].

2.5.4. Self-Cleaning

The blank and SiO$_2$NP/cotton/lycra composites were tested for self-cleaning by dyeing in a methylene blue dye solution (0.2 g/L) for 5 min and then drying it off. A one-half portion of each stain on the fabric was exposed to sunlight for 24 h to determine their color strength (K/S), whereas the other half portion was covered with black paper to prevent irradiation with sunlight [17]. The K/S value was evaluated using a Spectrophotometer CM-3600A. According to the following equation, decreasing K/S values refer to the level of dye degradation, which is stated as the self-cleaning capacity (SCC):

$$SCC = [(K/S)\,b - (K/S)\,a]/(K/S)\,b \times 100 \qquad (1)$$

where (K/S) a: color strength after exposing to daylight, and (K/S) b: color strength before.

3. Results and Discussion

A new method for developing high performance functional pile cotton fabric was designed. First, the functional performance of pile cotton fabrics was improved by incorporating a different percentage of lycra yarns with different densities according to the following parameters:

- **Weft material:** Cotton/lycra of six samples: three samples (90.8% cotton and 9.2% lycra), three samples (95.5% cotton and 5.5% lycra), and three cotton samples for the standard comparative.
- **Pile density:** 15, 19, 22 picks/cm.

Second, SiO$_2$ NPs were successfully synthesized and then applied on samples under study in a finishing bath containing an aqueous nano-SiO$_2$ dispersion and EDTA.

Statistical Analysis

All the data of antibacterial activity and self-cleaning were analyzed using descriptive statistics based on mean and standard deviation.

3.1. Characterization of Metal NPs Fabrics

3.1.1. SEM and EDX Analysis

The SEM images of both untreated cotton/lycra (95.5/5.5%) fabric of 15 pick/cm, cotton/lycra (90.8/9.2%) fabric of 19 pick/cm, and cotton (100%) fabric of 22 pick/cm and cotton/lycra (95.5/5.5%) fabric of 15 pick/cm, cotton/lycra (90.8/9.2%) fabric of 19 pick/cm, and cotton (100%) fabric of 22 pick/cm treated with SiO$_2$ NPs are shown in Figure 4A–F, respectively. The results indicated that SiO$_2$ NPs are well-distributed on the surface of the fabric, and the presence of SiO$_2$ NPs on the fabric surfaces was confirmed from EDX data. The elemental EDX data collected for the fabrics under study were presented in Figure 5. The peaks were allocated at 1.9 keV with 1.69% weight and 0.85% atomic absorption of the analyzed spot in the cotton (100%) fabric of 22 pick/cm, and at 1.9 keV with 1.08% weight and 0.54% atomic absorption of the analyzed spot in the cotton/lycra (95.5/5.5%) fabric of 15 pick/cm, as well as at 1.9 keV with 1.61% weight and 0.81% atomic absorption of the analyzed spot in the cotton/lycra (90.8/9.2%) fabric of

19 pick/cm fiber surface—characteristic of Si. The presence of these peaks confirmed that the SiO$_2$ NPs were entirely composed of SiO$_2$. The C and O signals originated from the cellulose polymer. Furthermore, the appearance of SiO$_2$ NPs on treated fabrics were also investigated by elemental mapping analysis, as illustrated in Figure 5E–G. It is apparent from the mapping images that silicon was distributed well on the surface confirming the uniform coating on cotton/lycra composites using SiO$_2$ NPs as a sufficient layer.

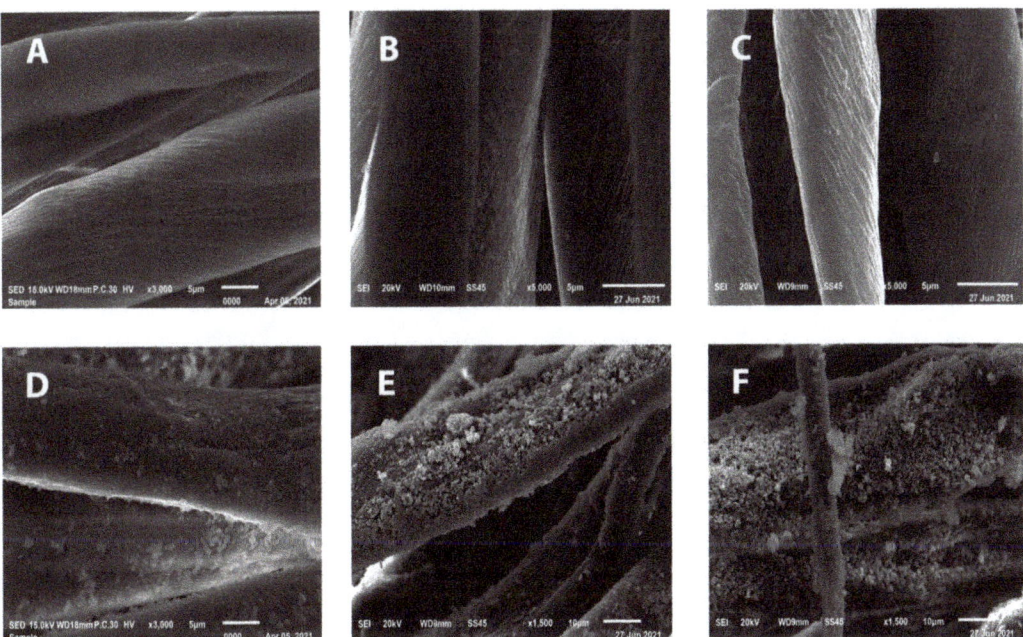

Figure 4. SEM images of (**A**–**C**) untreated samples. SEM images of cotton/lycra (95.5/5.5%) fabric of 15 pick/cm (**D**), cotton/lycra (90.8/9.2%) fabric of 19 pick/cm (**E**), and cotton (100%) fabric of 22 pick/cm (**F**) treated with SiO$_2$ NPs.

3.1.2. Antibacterial Activity of Blank and SiO$_2$-NP/Cotton/Lycra Composite Fabrics

The antibacterial properties of treated and untreated samples are summarized in Table 3. Ciprofloxacin is used as a standard to correlate the lead samples from the series (4A–I). It is an antibiotic, causing the production of oxidative radicals and bacterial cell death [36]. The results indicated that the untreated samples were not affected, and there were no inhibition areas. On the other hand, the SiO$_2$-NP/cotton/lycra composite fabrics exhibited excellent antibacterial activities against *S. aureus*, *Bacillus cereus*, *E. coli*, and *P. aeruginosa* compared with ciprofloxacin. This may be attributed to the fact that the mode of action of nanoparticles (NPs) is a direct interaction with the bacterial cell wall without the need to penetrate the cell. Most antibiotic resistance mechanisms can enhance the immunity of bacteria causing a lesser response to the antibacterial agent [37]. Samples E and G showed excellent activity against both Gram-positive and Gram-negative bacteria among all treated samples as compared to untreated samples. However, Sample (B) showed good activity against the four model bacteria. Samples (D) and (I) revealed good activity against both Gram-positive and Gram-negative bacteria in comparison to sample (A), which exhibited the lowest activity against Gram-negative (*P. aeruginosa*). Samples (F) and (H) showed good activity against the four model bacteria. However, sample (C) exhibited the lowest activity Gram-positive against (*Bacillus cereus*). In general, Samples (E) and (G) exhibited significantly high antibacterial activity compared to the all other samples in the series (4A–I) and with the standard. Additionally, the density of the pile 19 pick/cm is better than the other densities, as shown in Figure 6. This relies on the fact that there is more trapped air inside this structure (19 pick/cm) than in other densities. This trapped air helps in extending the zone of inhibition around the sample to 19 pick/cm, rendering it more resistant against bacterial attack. SiO$_2$ NPs' extraordinary antibacterial activity is attributed to their large surface area, which allows for better interaction with microbes [29].

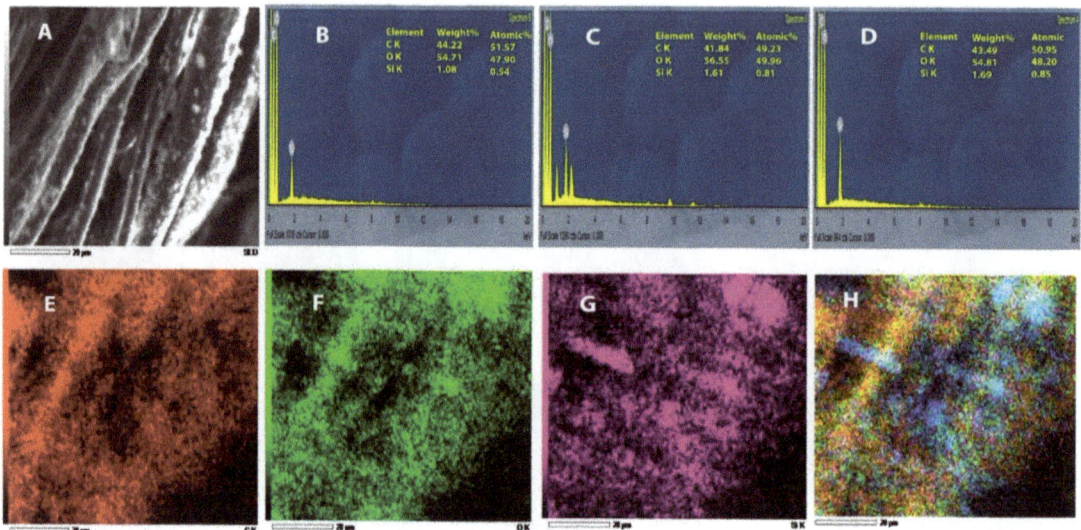

Figure 5. EDX spectrum of (**A**) SEM image (**B**) cotton/lycra (95.5/5.5%) fabric of 15 pick/cm, (**C**) cotton/lycra (90.8/9.2%) fabric of 19 pick/cm, and (**D**) cotton (100%) fabric of 22 pick/cm treated with SiO$_2$ NPs, and EDX mapping of cotton/lycra (90.8/9.2%) fabric of 19 pick/cm corresponding to (**E**) carbon, (**F**) oxygen, (**G**) silicon, and (**H**) overlap at 20 μm bar scale.

Table 3. Antibacterial Activity of the Blank and SiO$_2$NP/Cotton/Lycra Composites. Cotton (100%) 15 pick/cm (**A**), cotton (100%) 19 pick/cm (**B**), Cotton (100%) 22 pick/cm (**C**), cotton/lycra (90.8/9.2%) 15 pick/cm (**D**), cotton/lycra (90.8/9.2%) 19 pick/cm (**E**), cotton/lycra (90.8/9.2%) 22 pick/cm (**F**), cotton/lycra (94.5/5.5%) 19 pick/cm (**G**), cotton/lycra (94.5/5.5%) 22 pick/cm (**H**), cotton/lycra (94.5/5.5%) 15 pick/cm (**I**).

Type of NPs	Sample No.	Zone of Inhibition in mm				Mean	SD
		Gram Positive Bacteria		Gram Negative Bacteria			
		S. aureus	Bacillus cereus	E. coli	P. aeruginosa		
Untreated	1	0	0	0	0	0	0
Treated using SiO$_2$ NPs	A	22	13	21	5	15.25	7.93
	B	28	12	24	20	21	6.83
	C	28	13	27	28	24	7.35
	D	25	21	21	15	20.5	4.12
	E	30	30	30	26	29	2.00
	F	29	26	30	29	28.5	1.73
	G	29	27	31	31	29.5	1.91
	H	28	25	27	29	27.25	1.71
	I	19	17	23	22	20.25	2.75
Control Ciprofloxacin as antibiotic	J	21	22	26.7	23.3	23.25	2.49

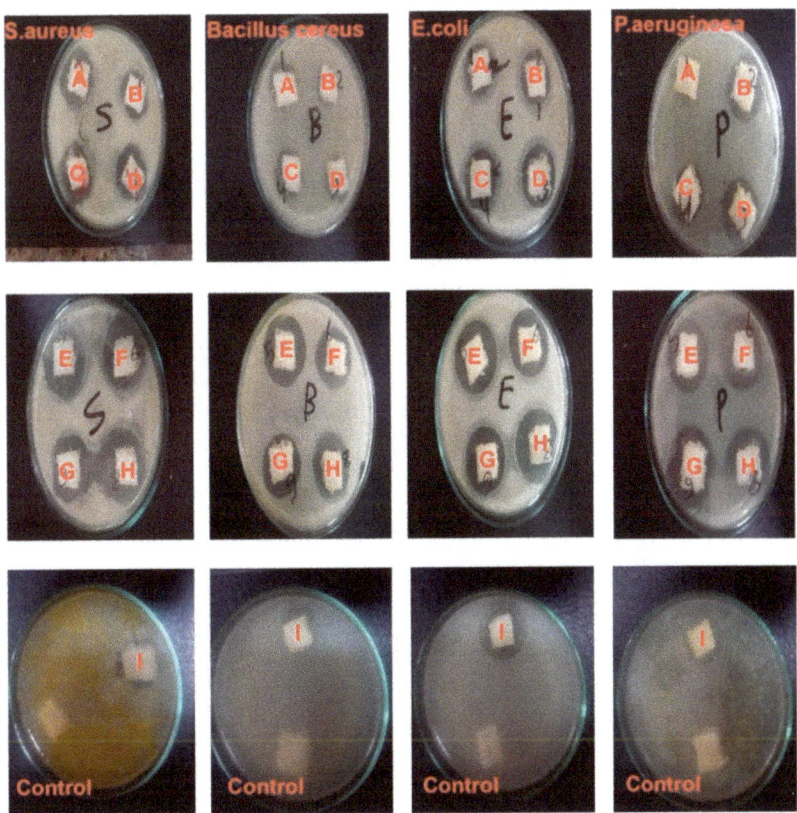

Figure 6. Photograph of Antibacterial Activity of the Blank and SiO$_2$ NP/Cotton/Lycra Composites.

3.1.3. UV Protection of Blank and SiO$_2$-NP/Cotton/Lycra Composite Fabrics

Table 4 showed the UPF values of the blank and SiO$_2$-NP/cotton/lycra composite fabrics. The results illustrated that both the treated and untreated samples have higher values of UPF compared with 100% cotton fabrics. The cover factor was expected to have a favorable impact on UV protection, since traditional cotton fabrics have a lower cover factor than sun protective woven or knitted materials. A main determining component of cover factor is the fabric construction parameter (ends/inch and picks/inch or courses/inch and wales/inch). Because threads are frequently interlaced, woven fabrics offer a larger cover factor than knitted fabrics, as shown in the cotton/lycra) composites under study. The pores between the yarns are small, allowing more radiation to penetrate. The threads in woven cotton/lycra composites are entirely opaque to UV radiation and the pores between the yarns are quite small. UV transmission is proportional to the porosity of the ideal fabric since light can only pass via the pores. The distances between pile fabrics, material thickness, and weight have an impact on the degree of protection. On the other hand, the higher weight and cover factor rate provide 95% of the best protection. The calculation of total UV percent transmittance for a fabric specimen is the ratio of the amount of radiation transmitted to the amount of radiation directed perpendicular to the fabric specimen surface. The cotton/lycra composites generally have no transmission when measuring. The term penetration or erythema weighted transmittance (EWT) is the inverse value of UPF Equation (2) and is frequently used to assign the degree of UVR protection of fabrics. EWT has a range of values between 0 and 1 (or 0 and 100%). The greater the sun protection provided by the fabric, the smaller the percentage of EWT. The results of EWT showed excellent protection (0.0002).

$$\text{EWT} = \frac{1}{\text{UPF}} \tag{2}$$

Table 4. UPF Values of the Blank and SiO$_2$ NP-Cotton/Lycra Composites.

Sample	UPF Values
100% Cotton	30
Untreated Cotton/Lycra (94.5/5.5%) 19 pick/cm	464.9
Treated Cotton/Lycra (94.5/5.5%) 19 pick/cm	2690.4
Untreated Cotton/Lycra (90.8/9.2%) 19 pick/cm	666.7
Treated Cotton/Lycra (90.8/9.2%) 19 pick/cm	4676.8

Note. UPF = UV protection factor.

The UPF values of SiO$_2$-NP-cotton/lycra composite fabrics exhibited excellent UV protection compared to those of blank fabrics, according to the Australian/New Zealand standard. Therefore, these results confirmed the UV refection ability of SiO$_2$ NPs, which can effectively decrease aging and reduce human skin damage caused by harmful UV radiation. However, the UPF value of the cotton/lycra (90.8/9.2%) 19 pick/cm is higher than that of the cotton/lycra (94.5/5.5%) 19 pick/cm sample. This may be the result of the finding that, the greater the percentage of lycra in the fabric, the greater the fabric cover factor and, consequently, the greater the UPF values.

3.1.4. Self-Cleaning of Blank and SiO$_2$-NP–Cotton/Lycra Composite Fabrics

Table 5 showed that the K/S values of SiO$_2$NP/cotton/lycra composite fabrics exposed to sunlight decreased compared to blank fabrics. Compared to the untreated samples, the results of both SiO$_2$-NP/cotton/lycra (94.5/5.5%) fabric of 19 pick/cm and cotton/lycra (90.8/9.2%) fabric of 19 pick/cm samples demonstrated greater self-cleaning activity, attributable to the fact that transition metal oxides and their composites, such as SiO$_2$ NPs, have strong photocatalytic activity for the photodegradation of organic pollutants, as illustrated in Figure 7. Metal oxide-based nanomaterials with well-controlled structural, crystalline, and surface characteristics behave as semiconductors with broadband gaps and have desirable attributes, including non-toxicity and stability.

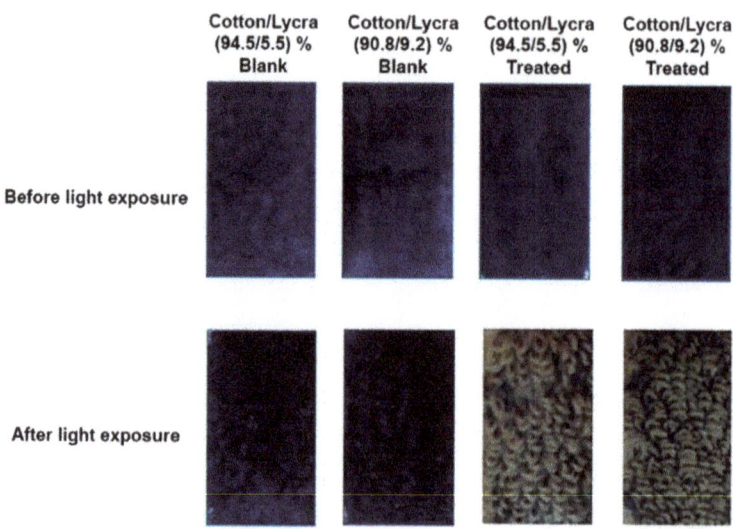

Figure 7. Self-cleaning of the blank and SiO$_2$NP/cotton/lycra composite fabrics before and after light exposure.

Table 5. Self-cleaning of the blank and SiO$_2$-NP/cotton/lycra composite fabrics.

Sample	K/S Value Unexposed	SD	K/S Value Exposed	SD	SCC
Blank	25.8	0.32	11.9	0.25	53.8
Cotton/Lycra (94.5/5.5%) 19 pick/cm	20.7	0.31	0.63	0.02	96.9
Blank	25.3	0.32	16.7	0.33	33.9
Cotton/Lycra (90.8/9.2%) 19 pick/cm	21.1	0.23	0.73	0.03	96.5

Note. K/S = (Color strength).

4. Conclusions

This research demonstrated the effect of both adding lycra yarns with different densities and implementation of SiO$_2$ NPs on the functional properties of plain pile cotton fabrics to produce composites. The SiO$_2$ NP coated cotton/lycra composites showed significant self-cleaning activity with 24 h of exposure. The self-cleaning results are identical in both the cotton/lycra (94.5/5.5%) and cotton/lycra (90.8/9.2%) 19 pick/cm samples (96.9 and 96.5 values), respectively. Besides the intrinsic antibacterial properties, the inclusion of SiO$_2$ NPs will simultaneously lead to the effective antibacterial activity of the fabricated surface. The pile density of 19 pick/cm provides higher bactericidal activity (30–31 mm) compared to other densities. The composite positively influenced the UV protection characteristics compared to blank samples. However, cotton/lycra (90.8/9.2%) 19 pick/cm sample showed higher UPF (4676.8) than did cotton/lycra (94.5/5.5%) 19 pick/cm sample (2690.4). In summary, the cotton/lycra (90.5/9.2%) 19 pick/cm sample showed the best results in terms of antibacterial activity, UV protection, and self-cleaning.

Author Contributions: Conceptualization, T.A.E. and G.M.E.; methodology, H.G.E.; software, G.M.E. validation, H.G.E., H.H.E. and T.A.E.; formal analysis, K.S.-A.; investigation, H.G.E.; resources, H.G.E.; data curation, H.G.E.; writing—original draft preparation, H.G.E.; writing—review and editing, T.A.E. and M.R.P.; visualization, M.R.P.; supervision, T.A.E.; project administration, T.A.E. All authors have read and agreed to the published version of the manuscript.

Funding: This research received no external funding.

Institutional Review Board Statement: Not applicable.

Informed Consent Statement: Not applicable.

Data Availability Statement: The data presented in this study are available on request from the corresponding author.

Conflicts of Interest: The authors declare no conflict of interest.

References

1. Toprak, T.; Anis, P. Textile industry's environmental effects and approaching cleaner production and sustainability, an overview. *J. Text. Eng. Fash. Technol.* **2017**, *2*, 429–442. [CrossRef]
2. Badr, A. Anti-microbial and durability characteristics of socks made of cotton and regenerated cellulosic. *Alex. Eng. J.* **2018**, *57*, 3367–3373. [CrossRef]
3. Ibrahim, H.M.; Zaghloul, S.; Hashem, M.; El-Shafei, A. A green approach to improve the antibacterial properties of cellulose based fabrics using Moringa oleifera extract presence of silver nanoparticles. *Cellulose* **2021**, *28*, 549–564. [CrossRef]
4. Mestin, S.; Sampath, V.R. Investigation on Mechanical and Comfort Properties of Denim Woven Fabrics Made from Different Spandex Percentage for Sports Wear. *Curr. Trends Fash. Technol. Text. Eng.* **2021**, *7*, 49–61. [CrossRef]
5. Alndijany, N. Evaluation of the comfort properties of pile fabrics. *Int. Des. J.* **2020**, *10*, 521–530. [CrossRef]
6. Tomar, R. *Handbook of Worsted Wool and Blended Suiting Process*; Woodhead Publishing India Pvt. Ltd.: New Delhi, India, 2010. [CrossRef]
7. Ahmed, M.; Morshed, M.; Farjana, S.; Ana, S. Fabrication of new multifunctional cotton–modal–recycled aramid blended protective textiles through deposition of a 3D-polymer coating: High fire retardant, water repellent and antibacterial properties. *New J. Chem.* **2020**, *44*, 12122–12133. [CrossRef]
8. Singha, K. Analysis of Spandex/Cotton Elastomeric Properties Spinning and Applications. *Int. J. Compos. Mater.* **2012**, *2*, 11–16. [CrossRef]
9. Kundu, S.; Chowdhary, U. Effect of fiber content on comfort properties of cotton/spandex, rayon/spandex, and polyester/spandex single jersey knitted fabrics. *Int. J. Polym. Text. Eng.* **2018**, *5*, 33–39. [CrossRef]

10. Prakash, C. Established the Effect of Loop Length on Dimensional Stability of Single Jersey Knitted Fabric Made from Cotton/Lycra Core Spun Yarn. *Indian J. Sci. Technol.* **2010**, *3*, 287–289. [CrossRef]
11. Abd-Elfattah, M.A.; Abou-Taleb, E.M.; Emara, A.I. Improving the performance of terry fabrics to be used as pilgrimage Clothes. *Int. Des. J.* **2020**, *10*, 397–413. [CrossRef]
12. Slater, K. Protection of or by textiles from environmental damage, in Environmental Impact of Textiles-Production, Processes and Protection, United Kingdom. *Text. Inst.* **2003**, 163–166. [CrossRef]
13. Beenu, S.; Manisha, G.; Rani, A. Development of UV Protective Clothing for College Going Girls. *Int. J. Curr. Microbiol. Appl. Sci.* **2019**, *8*, 1614–1621. [CrossRef]
14. Rildaa, Y.; Syukria, F.; Alifa, A.; Aziza, H.; Chandrenb, S.; Nurc, H. Self-cleaning TiO_2-SiO_2 clusters on cotton textiles prepared by dip-spin coating process. *J. Teknol. Sci. Eng.* **2016**, *78*, 113–120. [CrossRef]
15. Sobha, K.; Surendranath, K.; Meena, V.; Jwala, T.K.; Swetha, N.; Latha, K.S.M. Emerging trends in nanobiotechnology. *Biotechnol. Mol. Biol. Rev.* **2010**, *5*, 1–12. [CrossRef]
16. Abou Elmaaty, T.; Abdeldayem, S.; Ramadan, S.; Sayed-Ahmed, K.; Plutino, M. Coloration and Multi-Functionalization of Polypropylene Fabrics with Selenium Nanoparticles. *Polymers* **2021**, *13*, 2483. [CrossRef]
17. Ibrahim, N.; Eid, B.; Abd El-Aziz, E.; Abou Elmaaty, T.; Ramadan, S. Loading of chitosan Nano metal oxide hybrids onto cotton/polyester fabrics to impart permanent and effective multifunctions. *Int. J. Biol. Macromol.* **2017**, *105*, 769–776. [CrossRef]
18. Abou Elmaaty, T.; El-Nagare, K.; Raouf, S.; Abdelfattah, K.; El-Kadi, S.; Abdelazi, E. One-step green approach for functional printing and finishing of textiles using silver and gold NPs. *RSC Adv.* **2018**, *8*, 25546–25557. [CrossRef]
19. Cuk, N.; Sala, M.; Gorjanc, M. Development of antibacterial and UV protective cotton fabrics using plant food waste and alien invasive plant extracts as reducing agents for the in-situ synthesis of silver nanoparticles. *Cellulose* **2021**, *28*, 3215–3233. [CrossRef]
20. Attia, G.H.; Alyami, H.S.; Orabi, M.A.A.; Gaara, A.H.; El Raey, A.M. Antimicrobial activity of silver and zinc nanoparticles mediated by eggplant green calyx. *Int. J. Pharmacol.* **2020**, *16*, 236–243. [CrossRef]
21. Filipic, J.; Glazar, D.; Jerebic, S.; Kenda, D.; Modic, A.; Roskar, B.; Vrhovski, I.; Stular, D.; Golja, B.; Smolej, S.; et al. Tailoring of antibacterial and UV protective cotton fabric by an in-situ synthesis of silver particles in the presence of a sol–gel matrix and sumac leaf extract. *Tekstilec* **2020**, *63*, 4–13. [CrossRef]
22. Ahmad, S.; Fatma, I.; Manal, E.; Ghada, A.M. Applications of Nanotechnology and Advancements in Smart Wearable Textiles: An Overview. *Egypt. J. Chem.* **2020**, *63*, 2177–2184. [CrossRef]
23. Zhang, Y.Y.; Xu, Q.B.; Fu, F.Y.; Liu, X.D. Durable antimicrobial cotton textiles modified with inorganic nanoparticles. *Cellulose* **2016**, *23*, 2791–2808. [CrossRef]
24. Das, D.; Yang, Y.; Obrien, J.; Breznan, D.; Nimesh, S.; Bernatchez, S.; Hill, M.; Sayari, A.; Vincent, R.; Kumarathasan, P. Synthesis and physicochemical characterization of mesoporous SiO_2 nanoparticles. *J. Nanomater.* **2014**, *2014*, 62. [CrossRef]
25. Aya, R.; Abdel Rahim, A.; Amr, H.; El-Amir, M. Improving Performance and Functional Properties of Different Cotton Fabrics by Silicon Dioxide Nanoparticles. *J. Eng. Sci.* **2019**, *4*, 1–17. [CrossRef]
26. Bae, G.Y.; Jeong, Y.G.; Min, B.G. Superhydrophobic PET fabrics achieved by silica nanoparticles and water-repellent agent. *Fibers Polym.* **2010**, *11*, 976–981. [CrossRef]
27. Chakraborty, J.N.; Yashdeep, S. Comparative Performance of Synthesised Silica Nanoparticles for Enhanced Hydrophilic Properties on Cotton. *Tekstilec* **2019**, *62*, 278–287. [CrossRef]
28. Smiechowicz, E.; Niekraszewicz, B.; Strzelinska, M.; Zielecka, M. Antibacterial fibers containing nanosilica with immobilized silver nanoparticles. *AUTEX Res. J.* **2020**, *20*, 441–448. [CrossRef]
29. Riaz, S.; Ashraf, M.; Hussain, T.; Hussain, M.T. Modification of silica nanoparticles to develop highly durable superhydrophobic and antibacterial cotton fabrics. *Cellulose* **2019**, *26*, 5159–5175. [CrossRef]
30. Abbas, R.; Khereby, M.A.; Sadik, W.A.; El Demerdash, A.M. Fabrication of durable and cost effective superhydrophobic cotton textiles via simple one step process. *Cellulose* **2015**, *22*, 887–896. [CrossRef]
31. Maurice, A.R.; Faouzi, H. Synthesis and characterization of amorphous silica nanoparitcles from aqueous silicates using cationic surfactants. *J. Met. Mater. Miner.* **2014**, *24*, 37–42. [CrossRef]
32. Abou Elmaaty, T.; Elsisi, H.; Negm, E.; Ayad, S.; Sofan, M. Novel Nano Silica Assisted Synthesis of Azo pyrazole for the Sustainable Dyeing and Antimicrobial Finishing of Cotton Fabrics in Supercritical Carbon Dioxide. *J. Supercrit. Fluid* **2021**, in press.
33. Feifel, S.C.; Lisdat, F. Silica nanoparticles for the layer-by-layer assembly of fully electro-active cytochrome c multilayers. *J. Nanobiotechnol.* **2017**, 9–59. [CrossRef] [PubMed]
34. Ruiz-Cañas, M.C.; Corredor, L.M.; Quintero, H.I.; Manrique, E.; Bohórquez, A.R.R. Morphological and Structural Properties of Amino-Functionalized Fumed Nanosilica and Its Comparison with Nanoparticles Obtained by Modified Stöber Method. *Molecules* **2020**, *25*, 2868. [CrossRef]
35. Abou Elmaaty, T.; Kasem, A.; Mona Elsalamony, M.; Hanan, E. A Green Approach for One Step Dyeing and Finishing of Wool Fabric with Natural Pigment Extracted from Streptomyces Thinghirensis. *Egypt. J. Chem.* **2020**, *63*, 1999–2008. [CrossRef]
36. Masadeh, M.M.; Alzoubi, K.H.; Khabour, O.F.; Al-Azzam, S.I. Ciprofloxacin-induced antibacterial activity is attenuated by phosphodiesterase inhibitors. *Curr. Ther. Res.* **2014**, *77*, 14–17. [CrossRef]
37. Wang, L.; Chen Hu, C.; Shao, L. The antimicrobial activity of nanoparticles: Present situation and prospects for the future. *Int. J. Nanomed.* **2017**, *12*, 1227–1249. [CrossRef] [PubMed]

Article

One Surface Treatment, Multiple Possibilities: Broadening the Use-Potential of Para-Aramid Fibers with Mechanical Adhesion

Sarianna Palola [1,*], Farzin Javanshour [1], Shadi Kolahgar Azari [2], Vasileios Koutsos [2] and Essi Sarlin [1]

[1] Materials Science and Environmental Engineering Unit, Faculty of Engineering and Natural Sciences, Tampere University, FI-33014 Tampere, Finland; farzin.javanshour@tuni.fi (F.J.); essi.sarlin@tuni.fi (E.S.)
[2] School of Engineering, Institute for Materials and Processes, The University of Edinburgh, The King's Buildings, Robert Stevenson Road, Edinburgh EH9 3FB, UK; s.kolahgarazari@ed.ac.uk (S.K.A.); vasileios.koutsos@ed.ac.uk (V.K.)
* Correspondence: sarianna.palola@tuni.fi

Abstract: Aramid fibers are high-strength and high-modulus technical fibers used in protective clothing, such as bulletproof vests and helmets, as well as in industrial applications, such as tires and brake pads. However, their full potential is not currently utilized due to adhesion problems to matrix materials. In this paper, we study how the introduction of mechanical adhesion between aramid fibers and matrix material the affects adhesion properties of the fiber in both thermoplastic and thermoset matrix. A microwave-induced surface modification method is used to create nanostructures to the fiber surface and a high throughput microbond method is used to determine changes in interfacial shear strength with an epoxy (EP) and a polypropylene (PP) matrix. Additionally, Fourier transform infrared spectroscopy, atomic force microscopy, and scanning electron microscopy were used to evaluate the surface morphology of the fibers and differences in failure mechanism at the fiber-matrix interface. We were able to increase interfacial shear strength (IFSS) by 82 and 358%, in EP and PP matrix, respectively, due to increased surface roughness and mechanical adhesion. Also, aging studies were conducted to confirm that no changes in the adhesion properties would occur over time.

Keywords: aramid fibers; surface modification; adhesion; interphase; interfacial shear strength

1. Introduction

Para-aramid, poly (*p*-phenylene terephthalamide), fibers are highly crystalline synthetic fibers with high tensile strength, excellent chemical and abrasion resistance and high melting point. They even outrank carbon fiber in impact and wear resistance while also having higher a strength-to-weight ratio [1–3]. Aramid is used in fibrous form as well as in woven textile and pulp form, as reinforcement in demanding composite applications ranging from protective clothing (helmets, bulletproof vests, and fire protection) to automotive and industrial applications (gaskets, brake pads, tires, conveyor belts, and hoses).

However, the full use-potential of aramid fiber is hindered due to adhesion issues to matrix materials. To achieve the high strength-to-weight ratio, outstanding mechanical performance and durability characteristic of advanced composite materials [1–3], strong adhesion between the reinforcing fibers and the matrix material is critical. The adhesion issues with aramid fibers arise from the surface structure of the fiber, which is very smooth and chemically inert, lacking in reactive side groups [4,5]. To overcome this phenomenon, surface treatments are used, which traditionally promote either physical or chemical adhesion with the matrix. For example, a plasma treatment increases the surface energy of the fiber by increasing hydrogen bonds at the fiber surface, thus enabling a physical bond to be formed between the fiber and matrix [6–8]. On the other hand, with a chemical surface treatment, reactive side groups are grafted to the fiber surface, which can react with the matrix material and create a strong covalent bond between the fiber and the matrix [9,10]. However, these methods are often suitable for only one type of matrix material [11–13], may lose their effectiveness rapidly during storage [7] and may drastically reduce the mechanical properties of the fibers [14–16]. Thus, new approaches are needed, and research is increasingly directed towards utilizing mechanical adhesion between fibers and matrix [17].

Typically, mechanical adhesion is considered a lesser form of adhesion in composites, but it has some major advantages, such as independence of chemical compatibility. With mechanical adhesion or interlocking as the prominent adhesion mechanism, a wider range of material combinations could be used in composite applications, including thermoplastics. Thermoplastic materials are a desirable group of matrix materials for composite applications due to their lower toxicity and easier recyclability when compared to thermosetting materials. However, they are a challenging material group in terms of adhesion. Another benefit of mechanical adhesion at the fiber-matrix interface is that composite production becomes more economical as the same surface treatment can be used with multiple matrix types.

Mechanical adhesion or interlocking can be formed between the fiber and matrix, for example, by adding nanowires [18–20], nanoparticles [21–23], nanotubes [13,24–26], or nanofibers [27–29] to the fiber surface. These structures simultaneously increase the surface area and the surface roughness of the fiber. For example, by increasing mechanical adhesion together with chemical interactions, Nasser et al. [30] have been able to increase short beam strength of laser-induced graphene-coated aramid fabric by 70% in epoxy matrix. Lv et al. [13] have achieved similar results with in-situ polymer grafting and carbon nanotubes on aramid in the epoxy matrix, but they concluded the increase in interfacial shear strength (IFSS) to be due to increased polarity rather than topography. However, by purely increasing mechanical adhesion with adsorbed aramid nanofibers, Nasser et al. [27] have been able to increase short beam strength by 26% and IFSS by 70% in epoxy, which shows what the imminent potential mechanical adhesion has in terms of composite applications.

However, to fully benefit from mechanical adhesion, the attached medium (i.e., nanofibers or particles) needs also to be strongly adhered to the fiber surface, as Gonzalez-Chi et al. [24] and Ehlert et al. [20] have demonstrated. Also, the unique skin-core structure of the highly crystalline para-aramid fiber may lower the overall adhesion properties of the fiber even if strong interphase is formed between the fiber and matrix [31]. As force is applied to the interphase, the top layer of the fiber may fibrillate and be sheared off completely. By applying a "new skin" layer of graphene to the fiber, Cheng et al. [32] have been able to reconfigure the phenomenon and change the failure mechanism from fibrillation of the fiber "skin" to clean fracture at the interface, while increasing the IFSS by 75% in epoxy.

In this paper, we study the effect of mechanical adhesion as the main adhesion mechanism at the fiber-matrix interface. This is done by adding nanoscale deposits onto aramid fiber surface that increase surface area and topography and thus, enable mechanical adhesion at the fiber-matrix interphase. The concept of nanoscale deposit addition to increase adhesion in macroscale has been proven effective in our previous study [33]. However, the question remained whether the increased adhesion was purely due to mechanical adhesion or a combined effect (i.e., secondary entanglement) and would the result really be effective with other matrix materials as well. In this paper, we aim to address these questions and show that the effect is universal and does work with multiple matrix material types, and that the adhesion increase is purely due to increased mechanical adhesion. Also, we show that the effect is similar across different length scales ranging from micro to macroscale. Both thermoplastic and thermoset matrices were used to evaluate reliably the behavior of the nanodeposit decorated fiber surface in different matrix types, which have significantly different chemical and physical properties. Micromechanical testing is applied so that the failure mode and mechanism of the interphase can be monitored more closely and the effect of secondary artefacts, which may be present in macroscopic bulk material testing, such as fiber entanglement, can be eliminated from the results. For this, a high throughput microbond test system [34] was used to measure the IFSS of these nano-deposit decorated fibers. This test method was chosen over the more traditional fiber fragmentation test because fiber fragmentation test is unsuitable for aramid fibers due to their high-strain tensile failure mode [31]. Also, the microbond test method can be applied more easily to both thermoplastic and thermoset matrix materials. In order to focus on the effect of mechanical adhesion, polypropylene (PP) was chosen as the thermoplastic matrix material. PP has very limited hydrogen bonding interactions with the fiber surface, thus making it ideal for this type of investigation. Epoxy (EP) was used as the thermoset matrix because of its availability and wide use in polymer composites across the field. Fourier transform infrared spectroscopy (FTIR), scanning electron microscopy (SEM), and atomic force microscopy (AFM) were used to characterize the nanostructures, study the fiber-matrix interphase and identify the failure mechanism. Further, it was also investigated how well the widely debated microbond methodology represents macroscale properties of the composite by comparison to the previous results [33]. Also, the influence of aging during storage is studied, and what effect it has on the effectiveness of the surface treatment.

2. Materials and Methods

2.1. Aramid Surface Modification

Para-aramid fibers, Twaron 2201 (Teijin, Amsterdam, The Netherlands; properties according to the supplier: tensile strength 2.1 N/tex, linear density 1610 dtex, ~0.15 w-% sizing), were used as the fiber material. Prior to the surface treatment, the fibers were washed with mild detergent and rinsed with ethanol to remove the water-soluble, EO and PO alcohol component containing, surface sizing, and finally dried thoroughly. These fibers are denoted with suffix W, as 'washed'. Microwave-induced surface treatment was applied to a section of the washed fibers. This was done by first carburizing the fibers with Agar Turbo carbon evaporator B7230 (Agar Scientific, Stansted, UK) and then placing them into a glass container together with reactive chemicals (1:1 graphite and ferrocene). The container is then sealed and placed into a microwave oven, as described in the previous study [33]. Irradiation time of 14 s is used as it has been [33] determined to yield the best distribution and coverage of the fiber surface with the nanodeposits. These fibers are denoted with suffix MW. To investigate the storage properties of the MW—fibers, some of the fibers were kept in air at room temperature, protected from light, for 48 months prior to testing. Sample nomenclature is presented in Table 1.

Table 1. Sample nomenclature used in the study.

Sample Name	Matrix Material	Washing	Microwave Treatment
EP_W	epoxy	YES	NO
EP_MW	epoxy	YES	YES
PP_W	polypropylene	YES	NO
PP_MW	polypropylene	YES	YES

2.2. Interfacial Shear Strength

IFSS was measured with the microbond test method [35], in which single fiber microcomposite samples are prepared and tested. The samples are prepared by depositing droplets of the matrix material onto single fiber filaments and allowed to cure or cool down depending on the material type used. The droplets are then individually loaded with microvise blades until the droplet is detached from the fiber. To calculate the IFSS, the load required to detach the droplet (F_{max}) is then compared with the area of the fiber surface embedded by the droplet (A_{emb}), as described by Equation (1).

$$\frac{F_{max}}{A_{emb}} = \text{IFSS} \qquad (1)$$

For the IFSS testing in thermoset matrix, a low-viscosity epoxy (EP) resin system Araldite LY 5052/Aradur 5052 (Huntsman, Ratingen, Germany) was used, with a mixing ratio of 100/38, respectively. The resin was cured for 24 h at room temperature, followed by post-curing at 60 °C for 12 h. For the IFSS testing in thermoplastic matrix, a high melt flow heterophasic copolymer, polypropylene (PP) BJ380MO (Borealis AG, Vienna, Austria), was used.

Epoxy and PP droplets were dispensed onto the fibers with FIBROdrop (Fibrobotics Oy, Tampere, Finland) setup. A computer-controlled aluminum heating element was used to achieve optimum melt flow during PP droplet deposition. To prevent oxidation and thermal degradation of the PP melt during the droplet deposition, the FIBROdrop device was placed into an air-tight cabinet filled with nitrogen (N_2) gas. Also, a new batch of polymer melt was prepared for each fiber. High-throughput FIBRObond (Fibrobotics Oy, Tampere, Finland) [34] device was used for microbond measurements with a 1 N S-beam load cell and stainless steel sample holder. Testing was done in air at room temperature. Five fibers per sample type with 20–40 droplets per fiber were tested, resulting in approximately 100–200 data points for each sample type.

2.3. Microscopy

Field emission gun SEM Zeiss ULTRAplus (Zeiss, Oberkochen, Germany) was used for detailed imaging of the fiber surface and of the failed interface after IFSS testing. To minimize charging and to improve image quality, the samples were attached to aluminum sample holders with carbon glue and coated with carbon and tiny amount of gold.

Surface topography of the fibers was studied with an AFM MultiMode Nanoscope IIIa (Bruker, Santa Barbara, CA, USA). The measurements were done with tapping mode to gain information about possible phase shifts together with high spatial resolution while limiting the effect of artifacts and sample damage. Antimony (n) doped silicon tips of 0.01–0.025 Ohm-cm (model: NTESPA, Bruker, Santa Barbara, CA, USA) were used, which had a reflective aluminum coating on the backside. The imaging was done in air at room temperature. For the imaging process, samples were attached to a magnetic disc with double-sided adhesive. Data was analyzed with Gwyddion software [36].

2.4. FTIR Spectroscopy

The aramid fiber surface was analyzed with FTIR spectroscope Spectrum One (Perkin-Elmer, Buckinghamshire, UK). The Universal Attenuated Total Reflectance (ATR) sampling accessory of FTIR had a Diamond/ZeSe crystal with a 1.66 μm depth of penetration. Transmittance spectra were recorded within the 4000 to 600 cm^{-1} range and a 0.5 cm^{-1} resolution.

3. Results and Discussion

The fiber surface after the microwave irradiation treatment revealed an abundance of nanostructures covering the surface. As seen in Figure 1, the nanostructures are of irregular shape and that the topography of the fibers has changed due to the surface treatment significantly, but no visible voids are generated on the fiber surface. This is in line with our previous findings stating that the treatment has no negative effect on the mechanical properties of the fibers [33] and highlights the repeatability of the surface treatment method.

The AFM studies supported the SEM findings depicting clearly defined protrusions on the fiber surface. The phase contrast image highlights the structural and chemical difference between the bulk fiber and the nanostructures. As the color gradient in AFM phase contrast image is a combination of topographical details as well as changes in mechanical and adhesive properties, a contrast in color is created when the chemical and physical properties change in the imaged area. As the nanostructures appear brighter than the fiber surface, it can be deduced that they are not the same material as the fiber surface. Additionally, when using an Energy selective Backscattered (EsB) detector with SEM, the nanostructures also appear lighter than the bulk fiber itself, as seen in Figure 2.

The EsB detector reduces edge contrast in the image and thus, the apparent color difference between the bulk fiber and the nanostructures is due to increased Z-contrast [37] between the two. This, together with the AFM findings, means that the nanostructures are of different material and added to the surface during the microwave surface treatment rather than coming from the bulk fiber itself due to wrinkling or surface degradation.

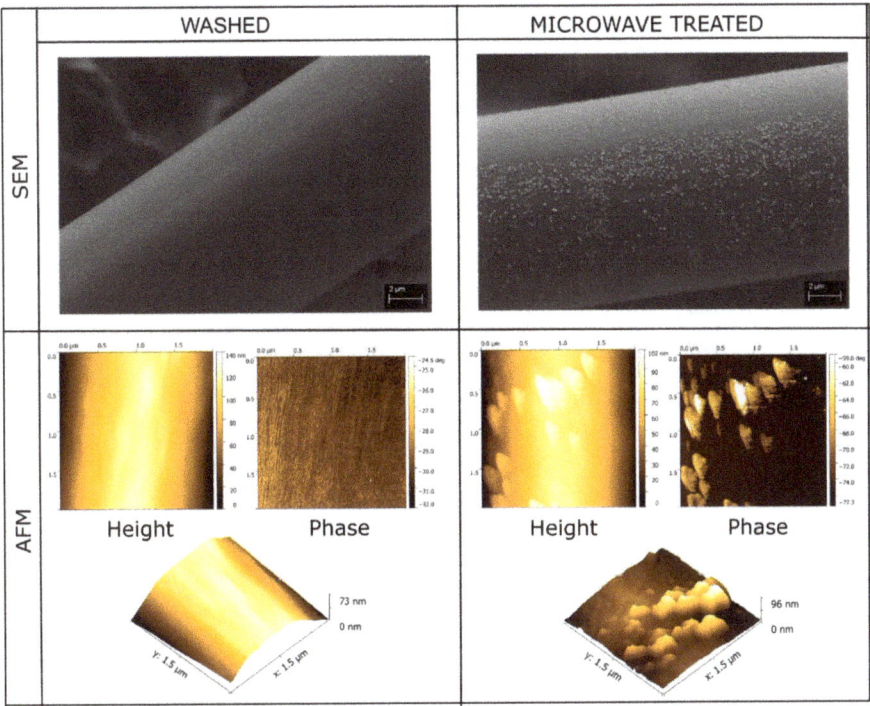

Figure 1. SEM images and corresponding AFM images of the aramid fibers before and after the microwave surface treatment.

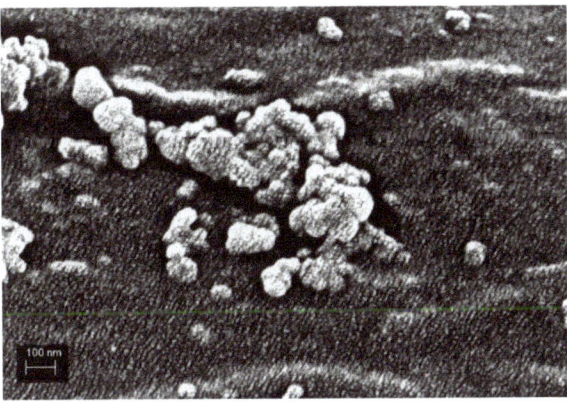

Figure 2. SEM EsB image of the microwave treated fiber surface highlighting the Z-contrast between the fiber surface and nanostructures.

The increased Z-contrast also implies that the nanostructures are mostly carbon-based compounds with traces of iron from ferrocene used in the microwave treatment. As a heavier element, iron would show up lighter in the EsB image, as seen in Figure 2. The iron molecule in ferrocene acts as a nucleation site for the carbon atoms as it is heated up during the treatment process [38] and thus can accumulate into the nanostructures. As seen from the figures, the irregular shape and varying size of the nanostructures increases the surface area of the fiber efficiently. This increases frictional forces at the fiber-matrix interface as well as adhesion through mechanical interlocking.

The FTIR spectrum (Figure 3) of the washed aramid fibers reveals characteristic peaks for para-aramid at 3312 cm^{-1} (–NH, hydrogen bond association states), 1637 cm^{-1} (C=O stretching vibration band of amide), 1537 cm^{-1} (N–H curved vibration), and 1305 cm^{-1} (N–H bending vibration) [39–41]. Compared to the FTIR spectrum of W-fibers, the hydrogen band peak of MW-fibers has broadened and moved to a lower wavenumber of 3305 cm^{-1} indicating increased hydrogen bonding at the surface and weakened hydrogen bonding in the polymer chains of the aramid fiber skin layer [40]. This means that the intense heat during the surface treatment causes some damage to the fiber surface but not to a degree that would affect the tensile properties of the fibers, as shown previously [33], or be visible in SEM. Also, a new peak is present at 2870 cm^{-1}, indicating CH_2/CH groups at the fiber surface [40]. The same peak is also present in ferrocene and graphite [42,43]. This confirms that the nanostructures are decomposition products of ferrocene and graphite, formed during the microwave irradiation treatment.

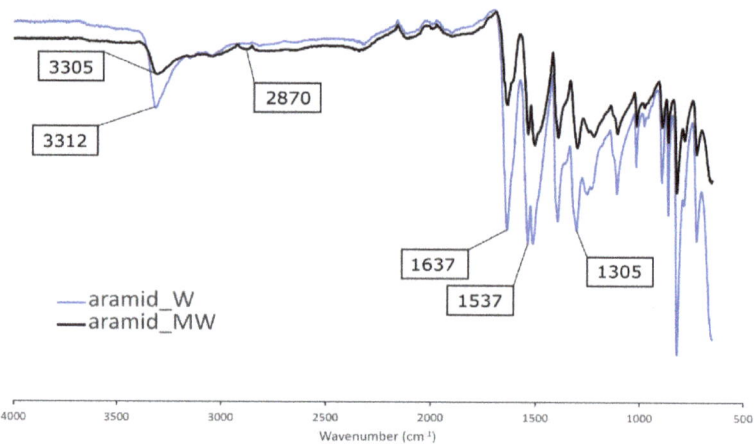

Figure 3. FTIR transmittance spectrum of washed (W) and microwave surface-treated (MW) aramid fibers.

IFSS was calculated with linear regression using the slope of load versus embedded area (Aemb) for each tested fiber separately. The IFSS for each sample type was then taken by calculating the average of the IFSS values of the separate fibers of that sample type. From the data in Figure 4, it can be seen that the load required to debond a droplet is higher with samples that are covered with nanostructures than with those that are not, even though the effective embedded area is similar. This implies that protrusions as small as nanoscale, can significantly alter the properties of the fiber-matrix interface in a way that can be detected with a microscale method. This same trend can be seen with both EP and PP matrix. As the behavior is similar in both thermoset and thermoplastic matrix, it emphasizes the importance of mechanical adhesion as a major adhesion mechanism that is independent of chemical compatibility. By increasing mechanical adhesion with the nanostructures, the maximum load increased by 56 and 395% in MW_EP and MW_PP, respectively. Although, the scattering of data appears to increase due to the surface treatment in MW_PP as compared to W_PP, this is not the case. Relative standard deviation (RSD) in both data sets is similar (~14%), which means that the data is highly comparable. Also, the R2 value for all measured samples ranged between 0.82–0.98, meaning high compatibility with the linear fit and thus, highly reliable measurement results.

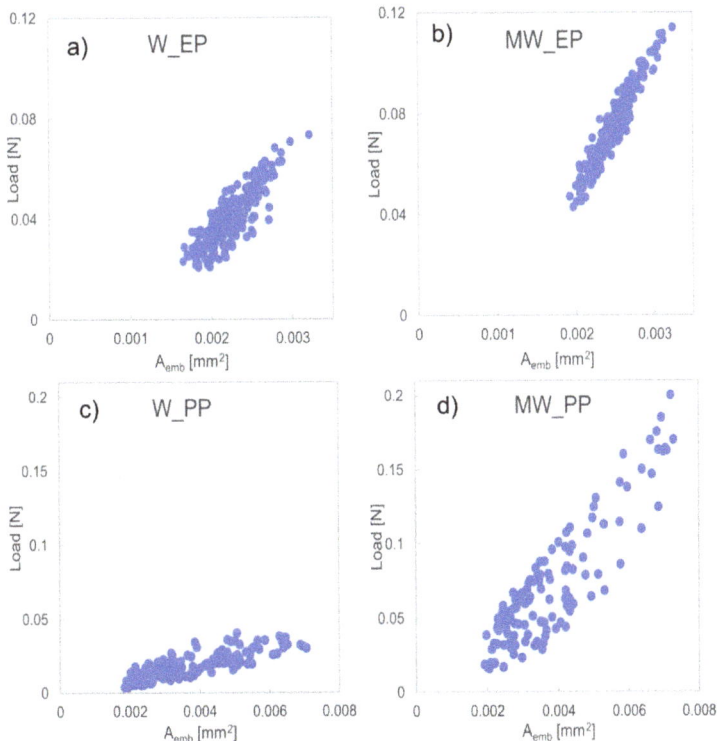

Figure 4. Load vs. embedded area for all measured droplets for (**a**) W_EP, (**b**) MW_EP, (**c**) W_PP, and (**d**) MW_PP.

The average IFSS results are presented in Figure 5 for W_EP, MW_EP, W_PP, and MW_PP together with macroscopic fiber bundle pull-out test results for the same surface treatment in rubber [33]. It is evident that the IFSS increases alongside with the increased surface topography of the fibers. Moreover, the increase in IFSS follows a similar trend with the bundle pull-out test in rubber. The results show that the IFSS increases in a similar fashion with thermoplastic, thermoset, and elastomeric matrices, even though the potential for chemical interaction of these matrix types is very different. For example, with EP, an increase in interfacial adhesion can be achieved through covalent bonding with the fiber surface during curing or by creating higher frictional force with the cured and cross-linked resin [13]. Mercaptan compounds, Lewis acid, and alkali products can be used to achieve such covalent bonds with EP. However, as none of them are grafted to the fiber surface in this case, what remains, is the increase in friction. This is also the case with PP. The chemical composition of PP provides only limited hydrogen interaction, which could affect favorably to interfacial adhesion with aramid. The main attribute towards the adhesion is mostly compressive forces due to favorable transcrystallization [31,39] occurring during the cooling process of the polymer melt. This was noted by Wang et al. [44]. They showed that small grooves and protrusions will increase thermal stress due to increased stress concentration during the PP crystallization when the melt is cooling. This localized stress concentration will further on enhance the nucleation ability of PP and, thus, promote transcrystallization leading to enhanced interfacial adhesion. The nanostructures created to the aramid fiber surface in this study, will act as such protrusions as described by Wang et al., and thus, lead to increased mechanical adhesion between the fiber surface and PP. In both EP and PP matrix, the nanostructures also increase stress transferability, which in turn, increases the IFSS in a similar fashion in both matrix types. These findings indicate that the primary adhesion mechanism between the fiber and polymer in this case, is indeed mechanical adhesion.

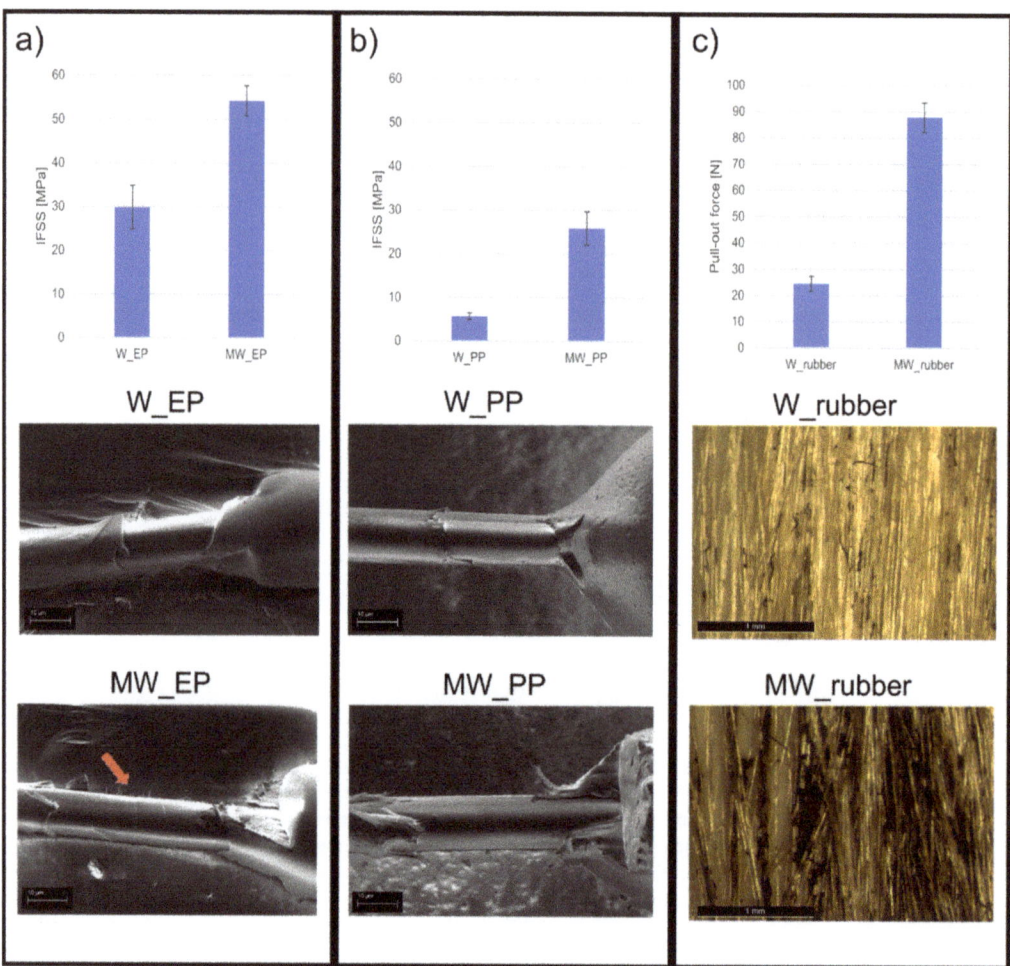

Figure 5. The IFSS results and SEM images of the corresponding fracture surfaces for (**a**) W_EP and MW_EP, where the arrow indicates fibrillation of the fiber surface, and (**b**) W_PP and MW_PP. The fiber bundle pull-out force and corresponding optical images are presented in (**c**) for W_rubber and MW_rubber [33].

The measured IFSS value for W_PP is 5.7 MPa, which is similar to other studies [24] done with microdroplet test and aramid/PP combination. This shows that the high-throughput microbond method is highly suitable for IFSS evaluation also with thermoplastic matrix. Overall, the IFSS increased from 29.8 to 54.2 MPa (82% increase) and from 5.7 to 25.9 MPa (358% increase), in EP and PP matrix, respectively, due to the surface treatment. This is very significant as it shows that the surface treatment is suitable for both thermoplastic and thermoset materials and that it has a similar effect in them both. Also, it is worthwhile noting that with this straightforward and fast surface treatment process, the IFSS of aramid/EP combination could be brought to the same level as with other more complicated methods reported only recently [32].

SEM images of the failed fiber-matrix interphase and visual observation during microbond testing supported the IFSS results. The failure mechanism during testing changed from pure shear at the interphase to a combination of peeling and shear as the surface topography was introduced (see videos in Supplementary Materials: Supporting Information files S1 and S2). With no surface treatment, the fiber surface after debonding appears smooth and unscathed, with only a minor amount of matrix residue remaining, as seen in Figure 5. This indicates that the fiber-matrix interphase has failed as the matrix droplet is sheared off. Also, the detachment site of the droplet shows a clean break with a small gap between the fiber surface and matrix, indicating a weak interphase. With the surface-treated fibers, the detachment site of the droplet shows no gap and is more uneven, as seen in Figure 6a,b, meaning that a stronger fiber-matrix interphase has been created. The debonded surface is rougher even with some fibrillation of the fiber skin structure, which indicates that the failure has shifted from purely occurring at the fiber-matrix interphase to a combination of fiber surface fibrillation and peeling together with interfacial shearing. The change in the appearance of the debonded surface is very clear with the harder EP matrix, as seen in Figures 5a and 6a, where red arrows point to sections of fibrillated fiber surface. Whereas with the softer PP, the matrix is rather fibrillating itself and clinging to the nanostructures than cleaving bits off from the fiber surface. The PP strands clinging to the fiber can be seen clearly underneath the fiber in Figure 6b. This indicates that the fiber/matrix adhesion is higher than the cohesive strength of the matrix. It also means, that the nanostructures are strongly attached to the fiber surface, as delamination occurs jointly from the skin-core interphase and skin-matrix interphase. As a result, it can be said that the main adhesion mechanism contributing towards the increased interfacial adhesion is mechanical adhesion.

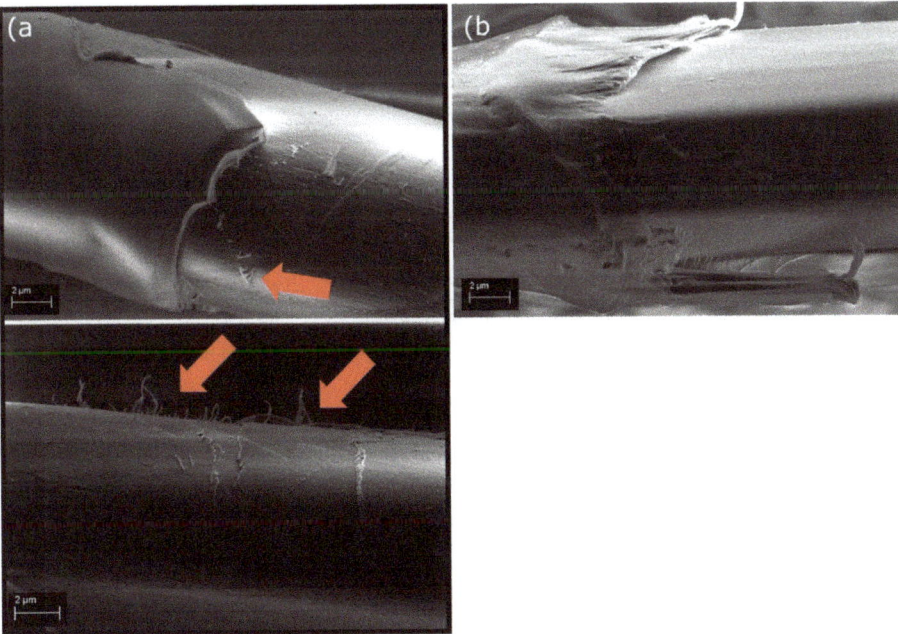

Figure 6. SEM images of the facture surfaces of (**a**) MW_EP, where red arrows indicate the fibrillation and peeling of the fiber surface, and (**b**) MW_PP after IFSS testing.

Additionally, the IFSS results follow very closely the same trend observed with the macroscopic fiber bundle pull-out test in rubber [33]. By increasing the amount of nanodeposits on the fiber surface, the adhesion and the strength of the interphase can be increased in both micro and macroscale. This is in line with findings of previous studies, such as the ones made by Beter et al. [45]. It also suggests that the high-throughput microbond method produces reliable data, which can indeed be used to evaluate adhesion properties in macroscopic composite structures. This type of composite research and development process can be made more economical and efficient.

Storage properties, and more precisely, aging, of the nanostructure covered fiber surface was also investigated. Some of the surface-treated fibers were taken aside and kept for 48 months at room temperature and protected from light. The IFSS of these fibers was measured with the microbond procedure in EP, and visual changes in the fiber surface were studied with SEM. The results revealed only minimal decrease in IFSS (~2%), which is well within the deviation range, compared to newly surface-treated fibers. Also, no change in the appearance of the fiber surface was observed. Thus, no significant decrease in the interfacial properties of the fibers has occurred, and the surface treatment can be considered durable enough to withstand storage over long periods of time.

4. Conclusions

This work explored the effect of nanostructures on the interfacial adhesion of aramid fiber in both a thermoplastic and a thermoset matrix and related the results also to an elastomeric matrix from a previous study. Our findings demonstrated that a significant increase in IFSS can be achieved in both thermoplastic (+358%) and thermoset (+82%) matrix, while maintaining mechanical and storage properties of the fibers. The increase in IFSS was noted to be due to enhanced mechanical adhesion between the fiber surface and matrix material caused by the addition of nanostructures to the fiber surface. The positive effect of the nanostructures on interfacial strength was observed both in micro and macroscale tests. The failure mechanism of the fiber-matrix interphase changes from clean shear to combined shear and peeling, as the level of mechanical adhesion increases, proving that the nanostructures are strongly attached to the fiber surface. These results highlight the significance of mechanical adhesion as the main adhesion mechanism and expand the use-potential of aramid fibers to multiple matrix material types and applications with just one fiber surface treatment.

Supplementary Materials: The following are available online at https://www.mdpi.com/article/10.3390/polym13183114/s1, Supporting Information file S1: A video of debonding of PP droplet on W_PP. Supporting Information file S2: A video of debonding of PP droplet on MW_PP.

Author Contributions: Conceptualization, S.P., V.K. and E.S.; Methodology, S.P.; Formal analysis, S.P.; Investigation, S.P., F.J., S.K.A. and E.S.; Data curation, S.P.; Writing—original draft preparation, S.P. and F.J.; Writing—review and editing, S.P., F.J., V.K. and E.S.; Supervision, V.K. and E.S.; Project administration, E.S.; Funding acquisition, S.P. All authors have read and agreed to the published version of the manuscript.

Funding: This research was funded by FINNISH CULTURAL FOUNDATION, grant number 00170813, TUTKIJAT MAAILMALLE and OTTO MALM FOUNDATION.

Institutional Review Board Statement: Not applicable.

Informed Consent Statement: Not applicable.

Data Availability Statement: The data presented in this study are available on request from the corresponding author.

Acknowledgments: Authors would like to thank Teijin Aramid Bv (Amsterdam, The Netherlands) for kindly providing the aramid fibers. This work made use of Tampere Microscopy Center facilities at Tampere University (33014 Tampere, Finland).

Conflicts of Interest: The authors declare no conflict of interest. The funders had no role in the design of the study; in the collection, analyses, or interpretation of data; in the writing of the manuscript, or in the decision to publish the results.

References

1. Karger-Kocsis, J.; Mahmood, H.; Pegoretti, A. Recent advances in fiber/matrix interphase engineering for polymer composites. *Prog. Mater. Sci.* **2015**, *73*, 1–43. [CrossRef]
2. Jesson, D.A.; Watts, J.F. The Interface and Interphase in Polymer Matrix Composites: Effect on Mechanical Properties and Methods for Identification. *Polym. Rev.* **2012**, *52*, 321–354. [CrossRef]
3. Tam, T.; Bhatnagar, A. High-performance ballistic fibers and tapes. In *Lightweight Ballistic Composites*, 2nd ed.; Bhatnagar, A., Ed.; Woodhead Publishing: Duxford, UK, 2016; pp. 1–39.
4. Chang, K.K.; Fibers, A. *ASM Handbooks*; Miracle, D.B., Donaldson, S.L., Eds.; Composites, ASM International: Materials Park, OH, USA, 2001; Volume 21, pp. 41–45.
5. Qi, G.; Zhang, B.; Du, S.; Yu, Y. Estimation of aramid fiber/epoxy interfacial properties by fiber bundle tests and multiscale modeling considering the fiber skin/core structure. *Compos. Struct.* **2017**, *167*, 1–10. [CrossRef]
6. Luo, S.; Ooij, W.J.V. Surface modification of textile fibers for improvement of adhesion to polymeric matrices: A review. *J. Adhes. Sci. Technol.* **2002**, *16*, 1715–1735. [CrossRef]
7. Sun, J.; Yao, L.; Sun, S.; Qiu, Y. ESR study of atmospheric pressure plasma jet irradiated aramid fibers. *Surf. Coat. Technol.* **2011**, *205*, 5312–5317. [CrossRef]
8. Wang, J.; Chen, P.; Xiong, X.; Jia, C.; Yu, Q.; Ma, K. Interface characteristic of aramid fiber reinforced poly(phthalazinone ether sulfone ketone) composite. *Surf. Interface Anal.* **2017**, *49*, 788–793. [CrossRef]
9. Tie-Min, L.; Yuan-Suo, Z.; Jie, H. Surface modification of Aramid fibers with new chemical method for improving interfacial bonding strength with epoxy resin. *J. Appl. Polym. Sci.* **2010**, *118*, 2541–2552.
10. Gao, B.; Zhang, R.; Gao, F.; He, M.; Wang, C.; Liu, L.; Zhao, L.; Cui, H. Interfacial Microstructure and Enhanced Mechanical Properties of Carbon Fiber Composites Caused by Growing Generation 1–4 Dendritic Poly(amidoamine) on a Fiber Surface. *Langmuir* **2016**, *32*, 8339–8349. [CrossRef] [PubMed]
11. Sa, R.; Yan, Y.; Wei, Z.; Zhang, L.; Wang, W.; Tian, M. Surface Modification of Aramid Fibers by Bio-Inspired Poly(dopamine) and Epoxy Functionalized Silane Grafting. *ACS Appl. Mater. Interfaces* **2014**, *6*, 21730–21738. [CrossRef] [PubMed]
12. Li, G.; Zhang, C.; Wang, Y.; Li, P.; Yu, Y.; Jia, X.; Liu, H.; Yang, X.; Xue, Z.; Ryu, S. Interface correlation and toughness matching of phosphoric acid functionalized Kevlar fiber and epoxy matrix for filament winding composites. *Compos. Sci. Technol.* **2008**, *68*, 3208–3214. [CrossRef]
13. Lv, J.; Cheng, Z.; Wu, H.; He, T.; Qin, J.; Liu, X. In-situ polymerization and covalent modification on aramid fiber surface via direct fluorination for interfacial enhancement. *Compos. Part B Eng.* **2020**, *182*, 107608. [CrossRef]
14. Yue, C.Y.; Padmanabhan, K. Interfacial studies on surface modified Kevlar fibre/epoxy matrix composites. *Compos. Part B Eng.* **1999**, *30*, 205–217. [CrossRef]
15. Zhang, S.; Li, M.; Cheng, K.; Lu, S. A facile method to prepare PEG coatings on the fiber surface by the reconstruction of hydrogen bonds for enhancing the interfacial strength of fibers and resins. *Colloids Surf. A Physicochem. Eng. Asp.* **2020**, *589*, 124426. [CrossRef]
16. Loureiro, L.; Carvalho, V.H.; Bettini, S.H.P. Reuse of p-aramid from industrial waste as reinforcement fiber in polyamide 6.6. *Polym. Test.* **2016**, *56*, 124–130. [CrossRef]
17. Palola, S.; Vuorinen, J.; Noordermeer, J.W.M.; Sarlin, E. Development in Additive Methods in Aramid Fiber Surface Modification to Increase Fiber-Matrix Adhesion: A Review. *Coatings* **2020**, *10*, 556. [CrossRef]
18. Hazarika, A.; Deka, B.K.; Kim, D.; Kong, K.; Park, Y.; Park, H.W. Growth of aligned ZnO nanorods on woven Kevlar® fiber and its performance in woven Kevlar® fiber/polyester composites. *Compos. Part A Appl. Sci. Manuf.* **2015**, *78*, 284–293. [CrossRef]
19. Malakooti, M.H.; Hwang, H.; Goulbourne, N.C.; Sodano, H.A. Role of ZnO nanowire arrays on the impact response of aramid fabrics. *Compos. Part B Eng.* **2017**, *127*, 222–231. [CrossRef]
20. Ehlert, G.J.; Sodano, H.A. Zinc oxide nanowire interphase for enhanced interfacial strength in lightweight polymer fiber composites. *ACS Appl. Mater. Interfaces* **2009**, *1*, 1827–1833. [CrossRef] [PubMed]
21. Patterson, B.A.; Sodano, H.A. Enhanced Interfacial Strength and UV Shielding of Aramid Fiber Composites through ZnO Nanoparticle Sizing. *ACS Appl. Mater. Interfaces* **2016**, *8*, 33963–33971. [CrossRef] [PubMed]
22. Cheng, Z.; Zhang, L.; Jiang, C.; Dai, Y.; Meng, C.; Luo, L.; Liu, X. Aramid fiber with excellent interfacial properties suitable for resin composite in a wide polarity range. *Chem. Eng. J.* **2018**, *347*, 483–492. [CrossRef]
23. Wang, B.; Duan, Y.; Zhang, J. Titanium dioxide nanoparticles-coated aramid fiber showing enhanced interfacial strength and UV resistance properties. *Mater. Des.* **2016**, *103*, 330–338. [CrossRef]
24. Gonzalez-Chi, P.I.; Rodríguez-Uicab, O.; Martin-Barrera, C.; Uribe-Calderon, J.; Canché-Escamilla, G.; Yazdani-Pedram, M.; May-Pat, A.; Avilés, F. Influence of aramid fiber treatment and carbon nanotubes on the interfacial strength of polypropylene hierarchical composites. *Compos. Part B Eng.* **2017**, *122*, 16–22. [CrossRef]
25. Hazarika, A.; Deka, B.K.; Kim, D.; Park, Y.; Park, H.W. Microwave-induced hierarchical iron-carbon nanotubes nanostructures anchored on polypyrrole/graphene oxide-grafted woven Kevlar® fiber. *Compos. Sci. Technol.* **2016**, *129*, 137–145. [CrossRef]

26. Chen, W.; Qian, X.; He, X.; Liu, Z.; Liu, J. Surface modification of Kevlar by grafting carbon nanotubes. *J. Appl. Polym. Sci.* **2011**, *123*, 1983–1990. [CrossRef]
27. Nasser, J.; Lin, J.; Steinke, K.; Sodano, H.A. Enhanced interfacial strength of aramid fiber reinforced composites through adsorbed aramid nanofiber coatings. *Compos. Sci. Technol.* **2019**, *174*, 125–133. [CrossRef]
28. Yang, M.; Cao, K.; Sui, L.; Qi, Y.; Zhu, J.; Waas, A.; Arruda, E.M.; Kieffer, J.; Thouless, M.D.; Kotov, N.A. Dispersions of Aramid Nanofibers: A New Nanoscale Building Block. *ACS Nano* **2011**, *5*, 6945–6954. [CrossRef] [PubMed]
29. Patterson, B.A.; Malakooti, M.H.; Lin, J.; Okorom, A.; Sodano, H.A. Aramid nanofibers for multiscale fiber reinforcement of polymer composites. *Compos. Sci. Technol.* **2018**, *161*, 92–99. [CrossRef]
30. Nasser, J.; Groo, L.; Zhang, L.; Sodano, H. Laser induced graphene fibers for multifunctional aramid fiber reinforced composite. *Carbon* **2020**, *158*, 146–156. [CrossRef]
31. Li, H.; Xu, Y.; Zhang, T.; Niu, K.; Wang, Y.; Zhao, Y.; Zhang, B. Interfacial adhesion and shear behaviors of aramid fiber/polyamide 6 composites under different thermal treatments. *Polym. Test.* **2020**, *81*, 106209. [CrossRef]
32. Cheng, Z.; Li, X.; Lv, J.; Liu, Y.; Liu, X. Constructing a new tear-resistant skin for aramid fiber to enhance composites interfacial performance based on the interfacial shear stability. *Appl. Surf. Sci.* **2021**, *544*, 148935. [CrossRef]
33. Palola, S.; Sarlin, E.; Azari, S.K.; Koutsos, V.; Vuorinen, J. Microwave induced hierarchical nanostructures on aramid fibers and their influence on adhesion properties in a rubber matrix. *Appl. Surf. Sci.* **2017**, *410*, 145–153. [CrossRef]
34. Laurikainen, P.; Kakkonen, M.; von Essen, M.; Tanhuanpää, O.; Kallio, P.; Sarlin, E. Identification and compensation of error sources in the microbond test utilising a reliable high-throughput device. *Compos. Part A Appl. Sci. Manuf.* **2020**, *137*, 105988. [CrossRef]
35. Miller, B.; Muri, P.; Rebenfel, L. A microbond method for determination of the shear strength of a fiber/resin interface. *Compos. Sci. Technol.* **1987**, *28*, 17–32. [CrossRef]
36. Nečas, D.; Klapetek, P. Gwyddion: An open-source software for SPM data analysis. *Cent. Eur. J. Phys.* **2012**, *10*, 181–188. [CrossRef]
37. *EsB Detector—Make Sub-Surface Information and Nano-Scale Composition Visible*; Carl Zeiss Microscopy GmbH: Oberkochen, Germany, 2016.
38. Bajpai, R.; Wagner, H.D. Fast growth of carbon nanotubes using a microwave oven. *Carbon* **2015**, *82*, 327–336. [CrossRef]
39. He, S.; Sun, G.X.; Cheng, X.D.; Dai, H.M.; Chen, X.F. Nanoporous SiO_2 grafted aramid fibers with low thermal conductivity. *Compos. Sci. Technol.* **2017**, *146*, 91–98. [CrossRef]
40. Li, Y.; Luo, Z.; Yang, L.; Li, X.; Xiang, K. Study on Surface Properties of Aramid Fiber Modified in Supercritical Carbon Dioxide by Glycidyl-POSS. *Polymers* **2018**, *11*, 700. [CrossRef]
41. Shebanov, S.M.; Novikov, I.K.; Pavlikov, A.V.; Anańin, O.B.; Gerasimov, I.A. IR and Raman Spectra of Modern Aramid Fibers. *Fibre Chem.* **2016**, *48*, 158–164. [CrossRef]
42. Mosadegh, M.; Mahdavi, H. Chemical Functionalization of Graphene Oxide by Poly(styrene sulfonate) Using Atom Transfer Radical and Free Radical Polymerization: A Comparative Study. *Polym. Plast. Technol. Eng.* **2017**, *56*, 1247–1258. [CrossRef]
43. Raja, P.M.V.; Barron, A.R. IR Spectroscopy. 2019. Available online: https://chem.libretexts.org/@go/page/167013 (accessed on 3 June 2021).
44. Wang, C.; Liu, C.R. Transcrystallization of polypropylene composites: Nucleating ability of fibres. *Polymer* **1999**, *40*, 289–298. [CrossRef]
45. Beter, J.; Schrittesser, B.; Fuchs, P.F. Investigation of adhesion properties in load coupling applications for flexible composites. *Mater. Today Proc.* **2020**, *34*, 41–46. [CrossRef]

Article

Influence of Structure and Composition of Woven Fabrics on the Conductivity of Flexography Printed Electronics

Ana María Rodes-Carbonell [1,*], Josué Ferri [2], Eduardo Garcia-Breijo [3], Ignacio Montava [4] and Eva Bou-Belda [4]

1. Textile Research Institut, Universitat Politècnica de València, 46022 Valencia, Spain
2. Instituto Tecnológico del Textil (AITEX), 03801 Alcoy, Spain; josue.ferri@aitex.es
3. Instituto Interuniversitario de Investigación de Reconocimiento Molecular y Desarrollo Tecnológico (IDM), Universitat Politècnica de València, 46022 Valencia, Spain; egarciab@eln.upv.es
4. Department of Textile and Paper Engineering, Universitat Politècnica de València, Plaza Ferrándiz y Carbonell s/n., 03801 Alcoy, Spain; imontava@txp.upv.es (I.M.); evbobel@upvnet.upv.es (E.B.-B.)
* Correspondence: anarodescarbonell@gmail.com

Citation: Rodes-Carbonell, A.M.; Ferri, J.; Garcia-Breijo, E.; Montava, I.; Bou-Belda, E. Influence of Structure and Composition of Woven Fabrics on the Conductivity of Flexography Printed Electronics. *Polymers* 2021, 13, 3165. https://doi.org/10.3390/polym13183165

Academic Editors: Tarek M. Abou Elmaaty and Maria Rosaria Plutino

Received: 28 August 2021
Accepted: 15 September 2021
Published: 18 September 2021

Publisher's Note: MDPI stays neutral with regard to jurisdictional claims in published maps and institutional affiliations.

Copyright: © 2021 by the authors. Licensee MDPI, Basel, Switzerland. This article is an open access article distributed under the terms and conditions of the Creative Commons Attribution (CC BY) license (https://creativecommons.org/licenses/by/4.0/).

Abstract: The work is framed within Printed Electronics, an emerging technology for the manufacture of electronic products. Among the different printing methods, the roll-to-roll flexography technique is used because it allows continuous manufacturing and high productivity at low cost. Nevertheless, the incorporation of the flexography printing technique in the textile field is still very recent due to technical barriers such as the porosity of the surface, the durability and the ability to withstand washing. By using the flexography printing technique and conductive inks, different printings were performed onto woven fabrics. Specifically, the study is focused on investigating the influence of the structure of the woven fabric with different weave construction, interlacing coefficient, yarn number and fabric density on the conductivity of the printing. In the same way, the influence of the weft composition was studied by a comparison of different materials (cotton, polyester, and wool). Optical, SEM, color fastness to wash, color measurement using reflection spectrophotometer and multi-meter analyses concluded that woven fabrics have a lower conductivity due to the ink expansion through the inner part of the textile. Regarding weft composition, cotton performs worse due to the moisture absorption capacity of cellulosic fiber. A solution for improving conductivity on printed electronic textiles would be pre-treatment of the surface substrates by applying different chemical compounds that increase the adhesion of the ink, avoiding its absorption.

Keywords: flexography; e-textiles; wearables; printed-electronics; textiles; electronic textiles

1. Introduction

Printed electronics (PE) refers to the technology that allows electronic device manufacturing through a printing process. PE is one of the fastest growing technologies in the world as it provides different printing techniques for manufacturing low-cost and large-area flexible electronic devices [1]. In recent years, flexible electronics technology has attracted considerable attention since it can be applied to wearable devices including flexible displays, flexible batteries and flexible sensors [2,3] in different areas such as aerospace and automotive, biomedical, robotics, and health applications [4]. Among them, wearable electronic textiles (e-textiles) are of great significance since they provide better comfortability, durability and lighter weight as well as maintaining desirable electrical properties [5].

The PE printing technique choice must be done according to the electronic application (e.g., small, thin, lightweight, flexible, and disposable, etc.), the manufacturing cost and volume. Additionally, the main materials (inks/pastes and substrates) must meet certain requirements, depending on the printing technology selected and the final application.

PE technologies are divided into contact techniques (e.g., flexography, gravure printing and soft lithography techniques), where the printing plate is in direct contact with the substrate and non-contact techniques (e.g., screen printing, aerosol printing, inkjet printing, laser direct writing), where only the ink contacts the substrate [6]. Those techniques suitable for roll-to-roll (R2R) processing, such as flexography, are especially attractive since they offer continuous manufacturing and high productivity [7].

Flexographic printing is known for depositing a wide range of thicknesses with the same resolution. The impression cylinder, plate cylinder, anilox roller, doctor blade and inking unit are the main parts of the flexographic printing process [1], as illustrated in Figure 1. Variables associated with the flexographic printing process include print speed, print force/engagement, anilox cell volume, anilox force/engagement as well as the ink and substrate properties [8]. Those variables have a direct impact on the print's morphological and electrical behavior. The print uniformity has a considerable influence on the final functionality of the device [9]. Within the context of printing electronics onto fabrics, it must be highlighted the challenge of durability and withstanding bending, stretching, abrasion and washing [10].

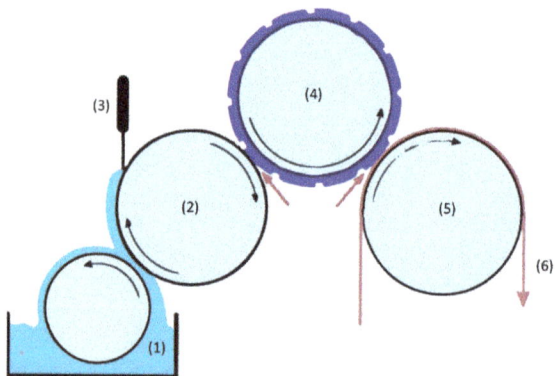

Figure 1. Significant parts of the flexography process: inking unit (1); anilox roller (2); doctor blade (3); plate cylinder (4); impression cylinder (5); printing substrate (6).

Numerous reviews and books were previously published, considering printed electronics on substrates such as glass, metal, paper or polymers [4,7]. Different studies were focused on the application of printed electronics onto textiles to obtain electrical devices such as capacitive sensors [11], perovskite solar cells [12] or RFID tags [13], but by using other printing techniques, such as ink jet or screen printing which are more common.

However, the incorporation of the flexography printing technique for printed electronics in the textile field is still very recent and there are not enough studies for its application. Continuing our previous work [14], this research is focused on the influence of the textile substrate parameters on the electrical performance of the printing.

The most used textile type for garments and even for technical applications is woven fabrics, where two sets of perpendicular yarns are crossed and interweave each other to create a coherent and stable structure [15]. In the context of this study, the term 'structure' refers to binding patterns of interlacing threads in woven fabrics, also considering internal structural features of the threads involved. Therefore, the structure of the woven fabric is determined, among other factors, by the weave construction, the interlacing coefficient, the density of threads in the fabric and the characteristics of warp and weft threads [16]. All the other physical (crimp, thickness, physical density, mass of unit area, porosity, etc.), mechanical (breaking strength and elongation, tensity, forces of rupture, resistance to abrasion, etc.) and permeability properties (permeabilities of gasses, liquids, light, sound, energy, water vapor, bacteria, etc.) are affected by previously mentioned selection [17]. Fabric texture and composition affect the porosity and strongly influence the textile characteristics such as the fabric mass, thickness, draping ability, stress–strain behavior, or air permeability. The surface topography of fabrics is responsible for their functionality—appearance and handle, wettability, soiling behavior and cleanability, abrasion resistance and wear. Topographical characteristics of the fabrics strongly depend on their construction parameters such as the type and fineness of filaments, yarn fineness, yarn density, and the type of weave. These characteristics have strong influence on, and in many cases, control the wetting properties [18].

It is important that printing ink has good adhesion to the substrate [19]. This becomes particularly complex in printing electronics onto fabrics where the intrinsic porous structures and texture characteristic of textiles affects the diffusion and penetration of conductive ink, being able to deteriorate the printing precision and electrical performance of conductive lines [20].

Related to the fabric structure, it was proved in other printing techniques, such as screen-printing, that the smallest pore size and roughness shows a higher printing precision and lower electrical resistance of printed conductive lines [20].

According to material composition, natural fibers such as cotton tend to absorb the ink more readily than synthetic fibers such as polyester, due to hydrophilic and wettability properties of cellulosic fibers [21]. Generally, synthetic textile materials have smooth, tight surfaces that offer little texture for ink adhesion [19].

However, there are not systematic investigations about the relationship between the electrical performance of a flexo printed textile with the structure of the woven fabric (i.e., interlacing coefficient, yarn count and weft density) as well as with the material composition of the weft.

With the aim of establishing this relation, different textiles were specifically manufactured varying their structural parameters and materials. After that, a controlled flexographic printing process was performed by using a silver electrical ink. Finally, fabrics were physically and electrically examined and compared.

The research and the obtained results are presented from two different approaches. On the one hand, an approach is followed from a perspective focused on distinguishing the impact of the structural parameters of the woven fabric on the conductivity of the printing. Structural parameters include, but are not limited to, weave construction, interlacing coefficient, yarn count and fabric density. On the other hand, from a point of view based on establishing the influence of the weft material composition on the conductivity, a comparison between fabrics with different weft materials (cotton, polyester, and wool) was also performed.

Physical and electrical analyses were carried out for both approaches. Several methodologies were used including optical, FE-SEM, color fastness to wash, and color measurement using reflection spectrophotometer or multi-meter analyses. Significant results were obtained and therefore studied.

It is believed that these findings will provide some important support for printing electronic devices on woven textiles by using the flexographic technique.

2. Materials and Methods

2.1. Materials

With the aim of establishing the influence of structure and composition of woven fabrics on the conductivity of flexographic printed electronics, eleven textile types were specifically manufactured for the research (Textiles Joyper S.L., Cocentaina, Spain), to be used as substrates for the printing.

On the one hand, eight different textiles structures were defined by varying the following constructional parameters: weave construction, weft yarn count, and weft density. On the other hand, textiles alternatives of the material used for the weft were added into the composition of three fabrics while maintaining the fabric structure.

Table 1 shows the main characteristics of these fabrics, where test samples from T1–T8 correspond to textile structure variations and textiles from T9–T11 to alternative weft materials. It should be noted that polyester yarn was used in the warp in all samples.

Table 1. Textile substrate's characterization (I): composition and ligament.

Substrate Code	3D Modeling	Weft Composition	Weave Construction	Graphic Representation	Interlacing Coefficient (KL)
T1		Polyester	Plain		1
T2		Polyester	Plain		1
T3		Polyester	Plain		1
T4		Polyester	Plain		1
T5		Polyester	Twill		0.4
T6		Polyester	Twill		0.4
T7		Polyester	Twill		0.4

Table 1. *Cont.*

Substrate Code	3D Modeling	Weft Composition	Weave Construction	Graphic Representation	Interlacing Coefficient (KL)
T8		Polyester	Twill		0.4
T9		Polyester	Twill		0.4
T10		Cotton	Twill		0.4
T11		Wool	Twill		0.4

The weave construction of textile materials is the regular structure produced by a pattern (unit cell) of interlaced threads repeating at regular intervals in two transversal directions [22]. The weave interlacing coefficient, KL, which depends on weave construction, is calculated by Equation (1):

$$KL = \frac{i}{w_1 \times w_2} \qquad (1)$$

where i is the number of interlacing points in the weave repeat, w_1 is the number of ends in the weave repeat, and w_2 is the number of picks in the weave repeat [23]. The research was focused on the two most traditional and commonly used weaves—plain and twill. The plain weave is the basic weave where one warp yarn is lifted over one weft yarn. The interlacing is opposite in all neighboring cells. Plain weave allows the highest possible number of interlacing. The twill weave has a pattern of diagonal lines; each warp yarn lifts over more than one weft, so the diagonal lines in fabric reach high densities [24].

Table 2 completes the main characteristics of the fabrics, adding the characteristics of the textile yarns in a detailed way for a better understanding.

Yarn number or yarn count refers to the thickness of a yarn and it is determined by its mass per unit length [25]. The fabric samples used in this study were produced by variation of the weft yarn count (167 dtex/333 dtex), without changing the warp count (167 dtex).

The density of the warp and weft is defined by the number of warps ends per cm and the number of picks per cm.

Table 2. Textile substrate's characterization (II): size and weight characteristics.

Substrate Code	Warp Density (ends/cm)	Weft Density (picks/cm)	Weft Count (dtex)	Cover Factor (%)	Fabric Weight [1] (g/m^2)	Thickness (μm)
T1	58.0	10.5	333.3	84.8	154	515
T2	58.8	15.3	333.3	93.6	171	550
T3	58.4	10.7	666.7	92.9	191	622
T4	57.2	15.8	666.7	103.4	233	650
T5	59.2	10.6	333.3	64.5	160	705
T6	55.9	16.0	333.3	68.3	180	725
T7	57.3	11.6	666.7	70.0	222	744
T8	56.3	16.5	666.7	77.6	241	805
T9	59.4	15.9	666.7	79.3	235	800
T10	58.8	16.2	666.7	78.5	230	795
T11	60.4	16.8	666.7	82.1	225	790

[1] Mass per unit area determined according to the standard ISO 3801.

Therefore, the structure variations among samples from T1–T8 are the following: the weave construction, as samples from T1–T4 are plain and T5–T8 are twill; the yarn count used in for the weft, which is thicker in samples T4, T5, T7 and T8; and the weft density, which is lower in samples T1, T3, T5 and T7.

Meanwhile, while maintaining the structure parameters, the difference among samples T9–T11 is the weft composition, which is made from polyester in T9, from cotton in T10 and from wool in T11.

For the research, other structural parameters have been accordingly obtained to complete the characterization of the woven fabric samples. Cover factor (the degree of fabric fullness) is the proportion of the fabric area covered by warp and weft yarns. It means that in practice, cover factor is calculated independently for warp and weft yarn by the proportion of fabric area covered by those yarn, according to Equation (2):

$$Cf = Cfwa + Cfwe - Cfwa \times Cfw \qquad (2)$$

as Cfwa = Dwa × dwa and Cfwe = Dwe × dwe, where Dwa and Dwe are densities of warp and weft and dwa and dwe are the diameter of warp and weft yarns, respectively. The cover factor directly depends on the yarn density and the yarn count [26]. Finally, the fabric weight was obtained according to the standard ISO 3801 by measuring the textile mass per unit area.

Regarding the ink, flexographic printing technology requires low-viscosity printing inks, which allows regular ink flow in the printing unit. Viscosity is generally lower than 0.05–0.5 Pa·s [27]. Same aqueous flexo-printable conductive ink, PFI-RSA6012—silver ink from Novacentrix (Austin, United States), was used in all prints to ensure comparable results. Details can be consulted in Table 3. The selection was made considering high conductivities and stretching properties for printing electronics on flexible substrates. It should be highlighted that there is a low supply on the market due to the novelty of the application of printed electronics in the textile through the flexography technique as stated above.

Table 3. PFI-RSA6012—Silver ink characteristics.

Ink Code	Density (g/mL)	Solids (%)	Viscosity (Pas)	Volume Resistivity (μΩ·cm)	Curing	Properties
PFI-RSA6012— Silver ink	2.22	60 (±2)	0.05–0.15 @1000 s^{-1}	8–10	10–60 s 140 °C	• Flexible • Compatible with Polyester

The ink contains silver nanoparticles and was formulated for high conductivity, fast curing, and improved levelling at lower printing speed.

2.2. Flexographic Electronic Printing

The manufacturing technology that is used is based on the flexographic printing technique of thick film. Flexography is a roll-to-roll direct printing technology, where an anilox roller, covered with micro-cavities on its surface, allows the collection of ink, and then is transferred to the printing plate cylinder. The specification of the anilox determines the volume of ink transferred to the printing plate. The ink is taken into these cells and the excess ink is subsequently removed by a doctor blade assembly.

For the research, one-layer flexographic prints were performed by using a printing experimental plant (K Printing Proofer, RK Print Coat Instruments Ltd., Litlington, United Kingdom). The equipment allows high quality proofs using flexography, among other printing techniques, with variable printing speeds of up to 40 m/min. Printing plates for use with the experimental plant are electronically engraved in exactly the same way as production cylinders. Using the flexo head, ink is transferred from the printing plate to a plain stereo roller and then onto the substrate. Adjustments can be made by micrometers. As developing a specific electronic device is out of the scope of this research, a plain design has been used for the test; nevertheless the printing plate and the printing roller in the industrial machine could be customized.

The experimental phase of this research consisted of the flexo-printing of the woven fabrics shown in Tables 1 and 2 using the silver ink described in Table 3. With the objective of studying the influence of the textile substrates on the conductivity of the printing, the equipment settings were kept fixed so that they did not interfere with the results. The specific setup conditions are shown in Table 4. Printed layers were dry cured in a FED-115 air oven from BINDER at 140 °C for one minute in order to use the same curing characteristics for all the samples. Moreover, previous to the printing, a thermic treatment was applied to all the fabrics to avoid variations of size due to the curing temperature of inks. The thermic treatment applied consisted of introducing the fabrics in the same oven at 130 °C for 15 min.

Table 4. Printing parameters.

Ink		Anilox Volume	Resolution	Printed Area	Speed	Curing
PFI RSA6012	Silver ink	11 cm^3/m^2	150 LPI	150 × 95 mm^2	12 m/min	60 s 140 °C

2.3. Characterization

Once dried, the printed textile samples were physically and electrically analyzed by carrying out several studies.

Regarding physical characterization, the following measurement methods were used: optical, scanning electron microscopy (FE-SEM) (Oxford Instruments plc, Abingdon, United Kingdom), and color fastness to wash and color measurement using reflection spectrophotometer.

The optical macroscopic images were taken with a LEICA MZ APO stereomicroscope. It was used to analyze the print uniformity of each layer of the electronic printed samples.

High resolution topographic images by SE (secondary electrons) and maps of crystalline and textural orientations by EBSD (electron backscatter diffraction) (Oxford Instruments plc, Abingdon, United Kingdom) were taken with a ZEISS ULTRA 55 Scanning Electron Microscope Field Emission Gun (field emission scanning electron microscopy (FE-SEM))) (Oxford Instruments plc, Abingdon, UK). They were used to analyze the ink penetration and the adhesion in each substrate.

Color fastness to domestic and commercial laundering was evaluated with a Gyrowash according to the standard method EN ISO 105-C06:2010. The test conditions were:

temperature of 40 °C, 10 steel balls and standardized ECE soap reference without optical or chemical whitener. After the test, the printed woven textiles were dried in a forced-air circulation dryer and treated samples were compared with untreated samples visually using a grey scale, according to ISO 105-A02 standard.

Color measurement was evaluated by the determination of CIELAB coordinates according to the standard method ISO 105-J01:1997. The apparatus used was DATACOLOR DC 650 (400–700 nm) (Datacolor, New Jersey, United States) with the following conditions: illuminant $D_{65}/10°$, diffuse measuring geometry and 6.6 mm of observation area.

With respect to electrical characterization, a usual two-terminal sensing unit was firstly considered to measure the conductivity behavior. Nevertheless, a two-wire system does not provide correct output due to variation in ambient temperature, as the resistance of the lead wires (both sides) changes unpredictably. Meanwhile, 4-wire Kelvin resistance measurement makes it possible to accurately measure resistance values less than 0.1 Ω while eliminating the inherent resistance of the lead wires connecting the measurement instrument to the component being measured [28]. For that reason, 4-wire system measurements were made. Resistance measurements were made with a FLUKE 8845A multimeter from FLUKE CORPORATION (Everett, WA, USA).

3. Results and Discussion

3.1. Electrical Characterization

For the purpose of determining the electrical conductivity of the printings an approach based on measuring the electrical resistance was considered. A low resistivity indicates a material that readily allows electric current. Table 5 shows a summary of the printing results for all the fabrics in terms of electrical resistivity. Four different samples of each fabric were measured. The orientation of the measurements on the woven fabrics were separately considered. In addition, results are graphically shown in Figure 2.

Table 5. Electrical resistance of the printings. Measurements have been made both in warp and weft directions.

Substrate Code	Resistance (Ω) Warp Direction	Resistance (Ω) Weft Direction
T1	0.4	0.3
T2	0.5	0.4
T3	0.4	0.5
T4	1.8	0.6
T5	0.4	1.1
T6	0.4	0.5
T7	0.5	0.4
T8	7	8.6
T9	7	8.6
T10	22.3	2.5
T11	0.3	0.4

Regarding the influence of the textile structure on the electrical behavior shown in Figure 2a, both measures on warp and weft directions agree that the woven fabric that presents the higher electrical resistance, and therefore worse electrical conductivity, is the substrate coded as T8. T8 was characterized by a lower interlacing coefficient and the highest mass per unit area.

These results coincide with the conclusions about the surface properties of woven fabric on electrical performance through a screen-printed technique. The fabric substrate with the smallest pore size and roughness shows a higher printing precision and lower electrical resistance of screen-printed conductive lines [20]. Therefore, the surface structure of the fabric substrate determines to some degree, not only the printing precision of conductive lines, but its electrical properties as well.

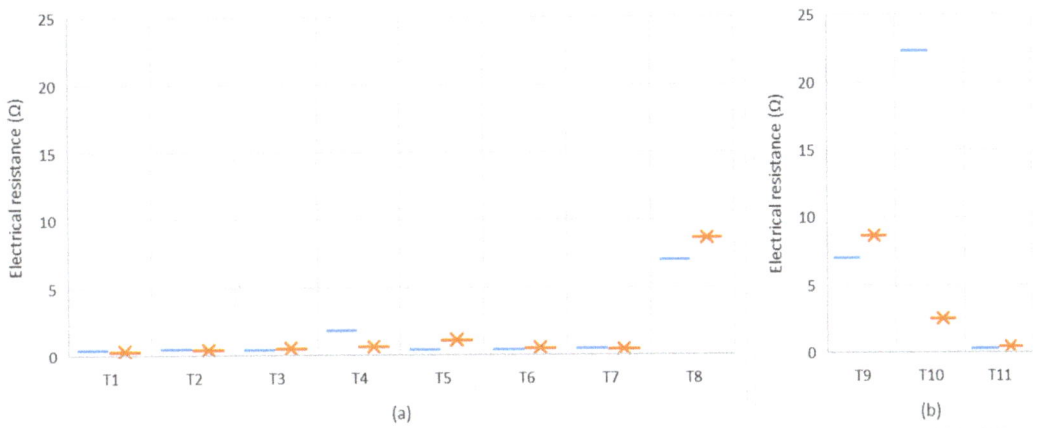

Figure 2. Graphic representation of the electrical resistance of the printings (—) measured on warp direction and (✖) measured on weft direction. (**a**) Electrical resistance affected by the structural variations of the textiles; and (**b**) Electrical resistance affected by the material composition of the textiles.

In order to address these challenges, surface pre-treatment onto rough and porous substrates or coatings and lamination processes should be performed in order to produce a continuous conductive pathway onto the textiles [5,10].

In order to deeply explore the relation found between woven fabric density and conductivity, results were plotted, as shown in Figure 3.

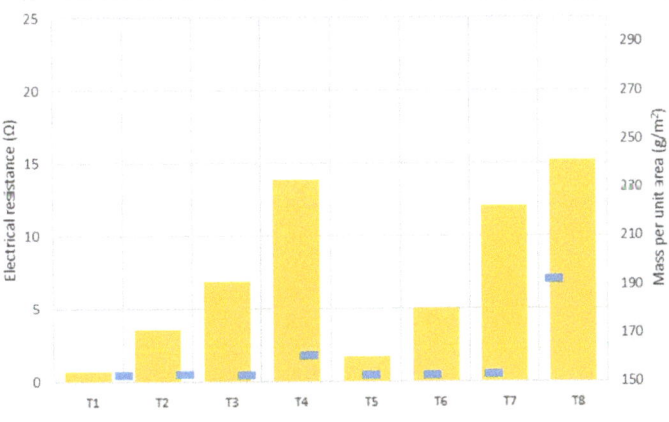

Figure 3. Graphic representation of the relation between electrical resistance of the printings measured on warp direction (—) and the mass per unit area (▪).

According to results shown in Figure 3, with regard to the influence of the textile structure on the electrical behavior, it could be expected that the higher the mass per unit, the lower conductivity due to the ink expansion through the inner part of the textile.

Moreover, Figure 2b allows the analysis in terms of the impact of the textile composition on the electrical behavior of electronic printing through the flexographic technique. Again, resistance measures in both directions agree and show that the woven fabric T10 presents higher electrical resistance. Even though the measure on the warp direction shows better electrical conductivity, it is worse than the other textile samples made from other materials. The weft of T10 was made from cotton. This makes sense, because of the moisture absorption capacity of cellulosic fiber. In this way it can be demonstrated that cotton performs worse as the weft material for printing electronics with the flexo technique.

An alternative solution for improving conductivity on printed electronic textiles with similar problems would be pre-treatment of the surface substrates by applying different chemical compounds that increase the adhesion of the ink avoiding its absorption.

3.2. Physical Characterization

Focusing on analyzing the print uniformity of each layer of the electronic flexo printed woven textiles, optical macroscopic images were taken with 20 magnifications. In parallel, in order to discuss the ink penetration and the adhesion in each substrate, high-resolution topographic images were taken with Scanning Electron Microscope Field Emission Gun (FE-SEM).

Table 6 shows the images obtained for the Optical and FE-SEM characterization of the woven substrates with different textile structures. The FE-SEM ($\times 100$) images show a visual characterization of the fabric and ink, and the FE-SEM ($\times 150$) images on the right show maps of crystalline and textural orientations by EBSD for a determination of the position of the silver particles.

Table 6. Optical and FE-SEM characterization (I).

Substrate Code	Optical ($\times 20$)	FE-SEM ($\times 100$)	FE-SEM ($\times 150$)
T1			
T2			
T3			

Table 6. Cont.

Substrate Code	Optical (×20)	FE-SEM (×100)	FE-SEM (×150)
T4			
T5			
T6			
T7			
T8			

Regarding optical results, images in Table 6 show a total color uniformity on the substrates' surface not depending on the textile structure. The silver shade obtained is typical from conductive inks which contains silver nanoparticles. In addition, the images reveal the porosity of the woven fabrics as they show their characteristic textiles holes, being different for each structure. It should be highlighted the picture of the substrate T8 as it is the one with more holes between the yarns forming the woven fabric, in comparison with the other samples.

In respect of FE-SEM results, thanks to images of the cross section of the substrates shown in Table 6, it can be observed how weft threads with different titles and densities are intertwined with the warp threads. These pictures reveal the enclaves formed by the interlacing of the threads in both directions, obtaining a roughness of the fabric and presenting different heights. Depending on the structure and yarn title, the enclave formed is different and it directly affects to the fabric thickness. As already stated on the fabrics' characterization in Table 2, the FE-SEM ($\times 100$) images in Table 6 confirm the thickness is greater for fabrics whose ligament is sarge (substrates T5–T8).

Furthermore FE-SEM ($\times 150$) results with more magnification allow to observe the ink penetration and distribution in the inner part of the textile. Images demonstrate that there are particles of the silver ink between the fibers of the weft threads.

Meanwhile, Table 7 shows the images obtained for the Optical and FE-SEM characterization of the woven substrates with different materials on the weft.

Table 7. Optical and FE-SEM characterization (II).

Substrate Code	Optical ($\times 20$)	FE-SEM ($\times 100$)	FE-SEM ($\times 150$)
T9			
T10			
T11			

Regarding ink uniformity, all the woven fabrics present a good visual level at the print in spite of the porous surface. To ensure this uniformity in printing electronics with low viscosity inks onto a rough and porous textile surface is a great challenge, due to the orientation of fibers or yarns and the change of fiber morphology constantly [5].

With respect to ink penetration, the images in Table 7 show that ink does not remain at the surface, but penetrates until the lower side of the textile. Thus, results coincide with the relation pointed out in Figure 3 about the influence of the mass per unit area on the electrical behavior. Whereas conventional printing that uses thickeners to increase the viscosity of the paste achieve deposition only on the surface, the low ink viscosity used in flexo printing penetrates to the inner of the textile, the interior of the textile. Thus, the topography of the sample plays an important role in achieving the homogeneity and continuity of the ink. Therefore, it is demonstrated that the higher the mass per unit, the lower conductivity due to the ink expansion through the inner part of the textile.

In terms of the impact of the material of the weft, the images shown in Table 7 do not show significant differences among them.

The color loss and staining resulting from desorption and/or abrasive action of the samples was evaluated according to EN ISO 105-C06:2010 tests for color fastness [29]. The grade of color fastness to domestic and commercial laundering is presented in Tables 8 and 9.

Table 8. Color fastness results (I).

Substrate Code	Change in Color	Staining					
		Wool	Acrylic	Polyester	Polyamide	Cotton	Acetate
T1	4–5	4–5	4–5	4–5	4–5	4–5	4–5
T2	4–5	4–5	4–5	4–5	4–5	4–5	4–5
T3	4–5	4–5	4–5	4–5	4–5	4–5	4–5
T4	4	4–5	4–5	4–5	4–5	4–5	4–5
T5	4–5	4–5	4–5	4–5	4–5	4–5	4–5
T6	4–5	4–5	4–5	4–5	4–5	4–5	4–5
T7	4–5	4–5	4–5	4–5	4–5	4–5	4–5
T8	4	4–5	4–5	4–5	4–5	4–5	4–5

Table 9. Color fastness results (II).

Substrate Code	Change in Color	Staining					
		Wool	Acrylic	Polyester	Polyamide	Cotton	Acetate
T9	4	4–5	4–5	4–5	4–5	4–5	4–5
T10	1	4	4	4	4	4	4
T11	1	4	4	4	4	4	4

For assessing the change in color, woven treated textiles were compared with a grey scale complying with ISO 105-A02. Regarding assessing staining, the grey scale used follows ISO 105-A03. Both scales consist of five pair of non-glossy grey color chips which illustrate the perceived color differences corresponding to fastness rating 5, 4, 3, 2 and 1. A range of 5 is given only when there is no perceived difference between the tested specimen and the original material.

According to results shown in Table 8, obtained values for change in color and staining are between 4 (Very Good) and 5 (Excellent). Therefore, it can be concluded that structural variations have not a relevant influence on the color fastness of the flexo electronic printings.

However, results shown in Table 9 demonstrate that color fastness decreases with natural fibers such as cotton and wool. The reason can be that ink adhesion improves in in synthetic fibers such as polyester thanks to the curing time after the flexography printing using 150 °C, as this type of fiber is a thermoplastic material and being its glass transition temperature around 70 °C [30].

In order to address these challenges in woven textiles with natural fibers, surface pretreatment should be done in order to improve the ink adhesion and therefore the electrical behavior [5]. It has been proved that special pretreatment processes on the fabric substrates improve the wash fastness for other printing techniques such as digital printing [31] or ink-jet [32]. According to a previous review [33], an increment of ink volume improves the ink coverage, upgrading in this case the conductivity, nevertheless it enhances the ink wash-out effect. For this reason, the ink volume transferred to the substrate should be optimized when conductivity and color fastness to washing are the objectives. In addition, coating and lamination processes could be done to ensure the continuous conductive pathway on textiles [10].

Spectrophotometric methods are adequate and objective for determining the color values of the fabric surface [34]. To assess variations found on the printings, measurements have been performed according to ISO 105-J01:1997 General principles for measurement of surface color [35].

The CIELAB, or CIE L* a* b*, color system represents quantitative relationship of colors on three axes: L*value indicates lightness, and a* and b* are chromaticity coordinates. It is the most widely used method for measuring and ordering object color. The results of the analysis of color of the printings onto the woven fabrics are shown in Table 10. For a better understanding, results have been plotted and can be consulted at the Figure 4. On the color space diagram, L* is represented on a vertical axis with values from 0 (black) to 100 (white). The a* value indicates red-green component of a color, where +a* (positive) and −a* (negative) indicate red and green values, respectively. The yellow and blue components are represented on the b* axis as +b* (positive) and −b* (negative) values, respectively. At the centre of the plane is neutral or achromatic [36].

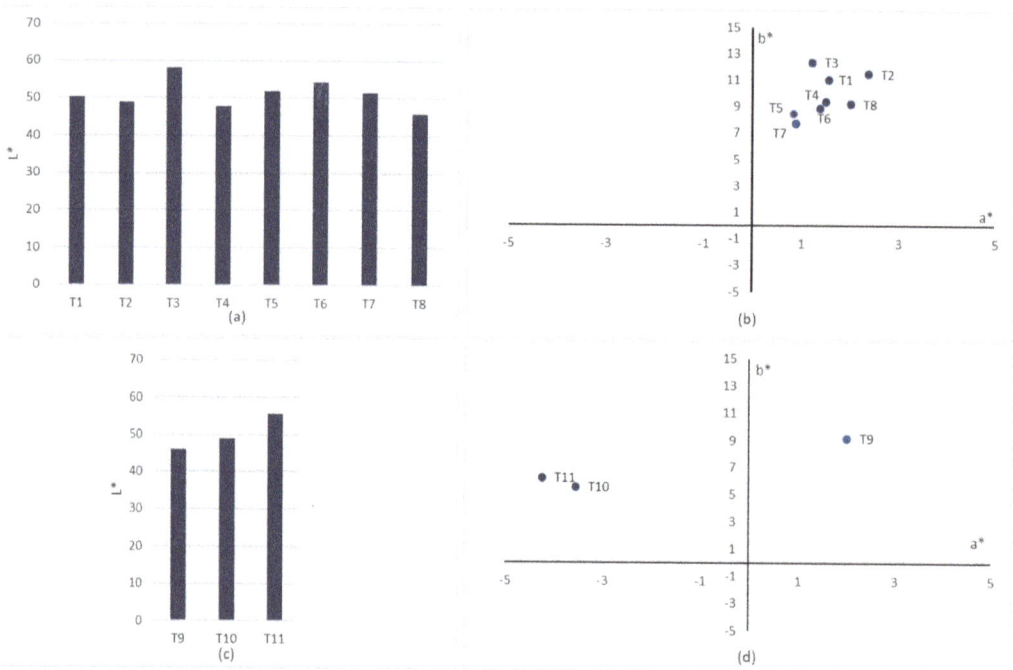

Figure 4. Two-dimensional CIELAB (+L*, +a *, +b*) coordinates representation. (**a**) CIELAB lightness (+L*) of textile structure variations T1–T8; (**b**) CIELAB a* and b* coordinates of textile structure variations T1–T8; (**c**) CIELAB lightness (+L*) of textile material variations T9–T11; (**d**) CIELAB a* and b* coordinates of textile material variations T9–T11.

Table 10. CIELAB* coordinates ($D_{65}/10°$).

Substrate Code	L*	a*	b*
T1	50.19	1.56	11.04
T2	48.86	2.36	11.48
T3	58.07	1.21	12.36
T4	47.78	1.49	9.36
T5	51.88	0.83	8.49
T6	54.29	1.37	8.86
T7	51.58	0.87	7.71
T8	45.95	2	9.19
T9	45.95	2	9.19
T10	48.91	−3.55	5.53
T11	55.58	−4.24	6.19

Lower L* indicates that the sample becomes darker, and all printed samples show low light (L*) values near to 50. It is observed that there is no correlation between the conductivity and the tone obtained from the coordinates a* and b* in contrast to silver coatings of polymeric substrates, where it is demonstrated that s shifting in the a* and b* during sintering means best conductivity [37]. On the other hand, it is observed that the sample T8/T9, the one with more weight and less conductivity, shows the lowest L* value, being darker than the other samples. This fact corroborates firstly, that variations on the fabric density have an influence on the color values [34], and secondly that the penetration of the silver particles into the textile, when the silver particles remain on the surface, thanks to emulsion resins, present values of L* between 60–70 [38].

4. Conclusions

It was proved that a direct relationship exists between the electrical performance of a flexographic printed textile with the structure of the woven as well as with the material composition of the weft.

After performing controlled flexographic printing processes onto woven fabrics by using a silver electrical ink, physical and electrical analysis led to the following conclusions. On the one hand, from a perspective focused on the structural parameters, the weave construction, interlacing coefficient, yarn count and fabric density, there was an effect due to the higher mass per unit of the fabric and lower conductivity due to the ink expansion through the inner part of the textile. On the other hand, based on the weft material composition (cotton, polyester, and wool), it was shown that cotton performs worse as weft material for printing electronics with the flexo technique due to the moisture absorption capacity of cellulosic fiber.

It is believed that these findings will provide some important support for printing electronic devices on woven textiles by using the flexographic technique.

The study's next steps will consist of improving the conductivity of flexographic printed textile by carrying out surface pre-treatment, coating and lamination processes that increase the ink adhesion and therefore the electrical behavior.

5. Patents

FERRI, J.; RODES-CARBONELL, A.M. and MORENO, J. (2020). Dispositivo NFC flexible para la medición, almacenamiento y transmisión de datos (Spain U202130440) http://www.oepm.es/pdf/ES/0000/000/01/26/48/ES-1264864_U.pdf (accessed on 12 April 2021).

Author Contributions: Conceptualization, J.F., E.G.-B., I.M., E.B.-B. and A.M.R.-C.; methodology, E.G.-B., I.M., E.B.-B. and A.M.R.-C.; validation, J.F., E.G.-B., I.M., E.B.-B. and A.M.R.-C.; formal analysis, A.M.R.-C.; investigation, J.F., E.G.-B., I.M., E.B.-B. and A.M.R.-C.; writing—original draft preparation, E.G.-B., I.M., E.B.-B. and A.M.R.-C.; writing—review and editing, E.G.-B., I.M., E.B.-B. and A.M.R.-C.; supervision, E.G.-B., I.M. and E.B.-B. All authors have read and agreed to the published version of the manuscript.

Funding: This work was supported by the Spanish Government/FEDER funds [Ministerio de Economía y Empresa (MINECO)/Fondo Europeo de Desarrollo Regional (FEDER)] under Grant RTI2018-100910-B-C43.

Institutional Review Board Statement: Not applicable.

Informed Consent Statement: Not applicable.

Data Availability Statement: The data presented in this study are available on request from the corresponding author.

Conflicts of Interest: The authors declare no conflict of interest. The funders had no role in the design of the study; in the collection, analyses, or interpretation of data; in the writing of the manuscript, or in the decision to publish the results.

References

1. Avuthu, S.G.; Gill, M.; Ghalib, N.; Sussman, M.R.; Wable, G.; Richstein, J.; Circuit, J. An introduction to the process of printed electronics. In Proceedings of the SMTA International, Rosemont, IL, USA, 25–29 September 2016.
2. Kim, C.; Jeon, S.W.; Kim, C.H. Reduction of Linearly Varying Term of Register Errors Using a Dancer System in Roll-to-Roll Printing Equipment for Printed Electronics. *Int. J. Precis. Eng. Manuf.* **2019**, *20*, 1485–1493. [CrossRef]
3. Wiklund, J.; Karakoç, A.; Palko, T.; Yiğitler, H.; Ruttik, K.; Jäntti, R.; Paltakari, J. A Review on Printed Electronics: Fabrication Methods, Inks, Substrates, Applications and Environmental Impacts. *J. Manuf. Mater. Process.* **2021**, *5*, 89. [CrossRef]
4. Rocha, L.A.; Viana, J.C. Printing Technologies on Flexible Substrates for Printed Electronics. 2018. Available online: https://doi.org/10.5772/intechopen.76161 (accessed on 15 June 2021). [CrossRef]
5. Karim, N.; Afroj, S.; Tan, S.; Novoselov, K.S.; Yeates, S.G. All Inkjet-Printed Graphene-Silver Composite Ink on Textiles for Highly Conductive Wearable Electronics Applications. *Sci. Rep.* **2019**, *9*, 8035. [CrossRef] [PubMed]
6. Cano-Raya, C.; Denchev, Z.Z.; Cruz, S.F.; Viana, J.C. Chemistry of solid metal-based inks and pastes for printed electronics—A review. *Appl. Mater. Today* **2019**, *15*, 416–430. [CrossRef]
7. Søndergaard, R.R.; Hösel, M.; Krebs, F. Roll to roll fabrication of large area functional organic materials. *J. Polym. Sci.* **2012**, *51*, 16–34. [CrossRef]
8. Mogg, B.T.; Claypole, T.; Deganello, D.; Phillips, C. Flexographic printing of ultra-thin semiconductor polymer layers. *Transl. Mater. Res.* **2016**, *3*, 015001. [CrossRef]
9. Davide, D. Control of morphological and electrical properties of flexographic printed electronics through tailored ink rheology. *Org. Electron.* **2019**, *73*, 212–218.
10. Yang, K.; Torah, R.; Wei, Y.; Beeby, S.; Tudor, J. Waterproof and durable screen printed silver conductive tracks on textiles. *Text. Res. J.* **2013**, *83*, 2023–2031. [CrossRef]
11. Ferri, J.; Llopis, R.L.; Moreno, J.; Lidón-Roger, J.V.; Garcia-Breijo, E. An investigation into the fabrication parameters of screen-printed capacitive sensors on e-textiles. *Text. Res. J.* **2020**, *90*, 1749–1769. [CrossRef]
12. Quinto, C.; Linares, A.; Llarena, E.; Montes, G.; González, O.; Molina, D.; Pío, A.; Ocaña, L.; Friend, M.; Cendagorta, M. Screen Printing for Perovskite Solar Cells Metallization. In Proceedings of the 31st European Photovoltaic Solar Energy Conference and Exhibition, Hamburg, Germany, 14–18 September 2015; pp. 1144–1148. [CrossRef]
13. Vena, A.; Perret, E.; Tedjini, S. Implementation and Measurements of Chipless RFID Tags. In *Chipless RFID Based on RF Encoding Particle*; ISTE Press—Elsevier: London, UK, 2016. [CrossRef]
14. Rodes-Carbonell, A.M.; Ferri, J.; Garcia-Breijo, E.; Bou-Belda, E. A preliminary study of printed electronics through flexography impression on flexible substrates. *Ind. Textila* **2021**, *72*, 133–137. [CrossRef]
15. Kumar, B.; Hu, J. Woven fabric structures and properties. In *The Textile Institute Book Series*; Woodhead Publishing: Sawston, UK, 2017; pp. 133–151. [CrossRef]
16. Amin, R.; Haque, M. Effect of weave structure on fabric properties. *Ann. Univ. Oradea Fascicle Text.* **2011**, *12*, 161–165.
17. Rogina-Car, B.; Kovačević, S.; Schwarz, I.; Dimitrovski, K. Microbial Barrier Properties of Cotton Fabric—Influence of Weave Architecture. *Polymers* **2020**, *12*, 1570. [CrossRef]
18. Hasan, M.; Calvimontes, A.; Synytska, A.; Dutschk, V. Effects of Topographic Structure on Wettability of Differently Woven Fabrics. *Text. Res. J.* **2008**, *78*, 996–1003. [CrossRef]
19. Meseldžija, M.; Vukić, N.; Erceg, T.; Budinski, N.; Lavicza, Z.; Lera, I.; Kojić, D. The analysis of the substrate influence on the print quality parameters of screen-printed textile. In Proceedings of the VIII International Conference on Social and Technological Development, Bosnia and Herzegovina, Balkans, 8–9 November 2019. [CrossRef]
20. Hong, H.; Hu, J.; Yan, X. Effect of the basic surface properties of woven lining fabric on printing precision and electrical performance of screen-printed conductive lines. *Text. Res. J.* **2019**, *90*, 1212–1223. [CrossRef]
21. Agathian, K.; Kannammal, L.; Meenarathi, B.; Kailash, S.; Anbarasan, R. Synthesis, characterization and adsorption behavior of cotton fiber based Schiff base. *Int. J. Biol. Macromol.* **2018**, *107*, 1102–1112. [CrossRef] [PubMed]
22. Grishanov, S.; Meshkov, V.; Omelchenko, A. A Topological Study of Textile Structures. Part I: An Introduction to Topological Methods. *Text. Res. J.* **2009**, *79*, 702–713. [CrossRef]

23. Özdemir, H.; Mert, E. The effects of fabric structural parameters on the tensile, bursting, and impact strengths of cellular woven fabrics. *J. Text. Inst.* **2013**, *104*, 330–338. [CrossRef]
24. Šajn Gorjanc, D. The Influence of Constructional Parameters on Deformability of Elastic Cotton Fabrics. *J. Eng. Fibers Fabr.* **2014**, *9*, 38–46. [CrossRef]
25. Vrljičak, Z.; Dragana, K.; Skenderi, Z. Impact of yarn count, noil percentage and yarn tension on structure of jersey fabric. *Tekstil* **2015**, *64*, 158–168.
26. Kostajnšek, K.; Dimitrovski, K. Use of Extended Cover Factor Theory in UV Protection of Woven Fabric. *Polymers* **2021**, *13*, 1188. [CrossRef] [PubMed]
27. Żołek-Tryznowska, Z. Rheology of Printing Inks. In *The Printing Ink Manual*; Springer: Berlin, Germany, 2016; pp. 87–99. [CrossRef]
28. Revuelta, P.S.; Litrán, S.P.; Thomas, J.P. Electrical Power Terms in the IEEE Std 1459 Framework. In *Active Power Line Conditioners*; Elsevier: Amsterdam, The Netherlands, 2016; pp. 23–49. [CrossRef]
29. Textiles—Tests for Colour Fastness—Part C06: Colour Fastness to Domestic and Commercial Laundering (ISO 105-C06:2010). Available online: https://www.une.org/encuentra-tu-norma/busca-tu-norma/norma?c=N0045649 (accessed on 15 June 2021).
30. Deopura, B.L.; Padaki, N.V. Synthetic textile fibres: Polyamide, polyester and aramid fibres. In *Textiles and Fashion*; Woodhead Publishing: Sawston, UK, 2015; pp. 97–114.
31. Jong, S.E.; Mok, L.Y.; Lee, J.S.; Chul, Y.S. Development of Cellulosic Woven Fabric for Digital Textile Printing. *Text. Coloration Finish.* **2005**, *17*, 20–26.
32. Zhang, Y.; Westland, S.; Cheung, V.; Burkinshaw, S.M.; Blackburn, R.S. A custom ink-jet printing system using a novel pretreatment method. *Color. Technol.* **2009**, *125*, 357–365. [CrossRef]
33. Kašiković, N.; Vladić, G.; Milić, N.; Novaković, D.; Milošević, R.; Dedijer, S. Colour fastness to washing of multi-layered digital prints on textile materials. *J. Natl. Sci. Found. Sri Lanka* **2018**, *46*, 381. [CrossRef]
34. Gabrijelčič, H.; Dimitrovski, K. Influence of yarn count and warp and weft thread density on colour values of woven surface. *Fibres Text. East. Eur.* **2004**, *12*, 32–39.
35. Available online: https://www.une.org/encuentra-tu-norma/busca-tu-norma/norma?c=N0023384 (accessed on 15 July 2021).
36. Ly, B.C.K.; Dyer, E.B.; Feig, J.L.; Chien, A.L.; Del Bino, S. Research Techniques Made Simple: Cutaneous Colorimetry: A Reliable Technique for Objective Skin Color Measurement. *J. Investig. Dermatol.* **2020**, *140*, 3–12.e1. [CrossRef] [PubMed]
37. Cherrington, M.; Claypole, T.; Gethin, D.; Worsley, D.; Deganello, D. Non-contact assessment of electrical performance for rapidly sintered nanoparticle silver coatings through colorimetry. *Thin Solid Films* **2012**, *522*, 412–414. [CrossRef]
38. Soleimani-gorgani, A.; Pishvaei, M. Water Fast Ink Jet Print Using an Acrylic /Nano-Silver Ink. *Prog. Color Colorants Coat.* **2021**, *4*, 79–83. [CrossRef]

Article

Preparation of Multifunctional Plasma Cured Cellulose Fibers Coated with Photo-Induced Nanocomposite toward Self-Cleaning and Antibacterial Textiles

Hany El-Hamshary [1,2,*], Mehrez E. El-Naggar [3,*], Tawfik A. Khattab [3] and Ayman El-Faham [1,4]

1. Chemistry Department, College of Science, King Saud University, Riyadh 11451, Saudi Arabia; aelfaham@ksu.edu.sa
2. Department of Chemistry, Faculty of Science, Tanta University, Tanta 31527, Egypt
3. Textile Research Division, National Research Center (Affiliation ID: 60014618), Cairo 12622, Egypt; ta.khattab@nrc.sci.eg
4. Department of Chemistry, Faculty of Science, Alexandria University, Ibrahimia, Alexandria 21321, Egypt
* Correspondence: helhamshary@ksu.edu.sa (H.E.-H.); mehrez_chem@yahoo.com (M.E.E.-N.)

Abstract: Multifunctional fibrous surfaces with ultraviolet protection, self-cleaning, or antibacterial activity have been highly attractive. Nanocomposites consisting of silver (AgNPs) and titanium dioxide (TiO_2 NPs) nanoparticles (Ag/TiO_2) were developed and coated onto the surface of viscose fibers employing a straightforward pad–dry–cure procedure. The morphologies and elemental compositions were evaluated by scan electron microscopy (SEM), infrared spectra (FTIR), and energy-dispersion X-ray spectra (EDS). The resultant multifunctional textile materials displayed antibacterial and photo-induced catalytic properties. The photocatalyzed self-cleaning properties were investigated employing the photochemical decay of methylthioninium chloride, whereas the antibacterial properties were studied versus *E. coli*. The viscose fibers coated with Ag/TiO_2 nanocomposite demonstrated improved efficiency compared with viscose fibers coated with pure anatase TiO_2 nano-scaled particles.

Keywords: Ag/TiO_2; nanocomposite; antibacterial; photocatalysis; viscose fibers

1. Introduction

High-performance textiles have significant potential in marketing nano-based functional commodities, such as antibacterial and self-cleaning textiles [1–5]. These nano-based functional textiles can be accomplished by the immobilization of metal and/or metal oxide nanoparticles onto the fabric surface during the finishing process. The great prospects of metal nanoparticles can be effectively employed to provide multifunctional stimulation without deteriorating the exterior properties or negatively affecting the native features of the fibers [6]. Various studies were explored for the usage of Ag^0, TiO_2, and ZnO nanoparticles as agents for textile surface modification to provide smart fibers with a variety of distinctive properties, like ultraviolet blocking, self-cleaning, and antibacterial activity [7–9]. Different techniques were described recently to tie TiO_2 nanoparticles into the fiber surface to present self-cleanable products [10]. The photocatalytic performance of TiO_2 nanoparticles upon irradiation with a UN supply was described. The exposure of TiO_2 nanoparticles to UV ($\lambda < 388$ nm) results in stimulating the electrons of the valence band into the other conduction one to generate holes (h^+) and electrons (e^-). Those reactive entities showed a major role in the commencement of a reduction-oxidation course [11]. TiO_2 nanoparticles have been reported as a high-quality substance in photocatalysis under irradiation with an ultraviolet supply owing to its satisfactory optical properties, chemical/physical stability, non-toxicity, and cheapness [12,13]. Nonetheless, some weakness was linked to the use of TiO_2 nanoparticles, such as an elevated band-gap (Eg = 3.2 eV). In addition, TiO_2 nanoparticles can be excited only under irradiation with an ultraviolet supply (λ < 388 nm) to release electrons to conduction band departing holes to the other valence one, limiting their photocatalytic activity under visible or sunlight. Furthermore, the high recombination rate between holes and electrons on TiO_2 nanoparticles results in less effective photocatalysis [2–5].

Silver nanoparticles (AgNPs) have been applied as an antimicrobial agent onto a variety of textile substrates in the absence of UV light [14,15]. However, silver nanoparticles can simply influence the colorimetric properties of the treated textile surface by oxidation into the brownish AgO or by aggregation into bigger black microparticles. In addition, silver is a costly metal, and small amounts are ineffective for a variety of realistic products. In order to accomplish the advantageous effects from both Ag^0 and TiO_2 nanoparticles and reduce their weaknesses, Ag/TiO_2 composites were developed by producing AgNPs onto TiO_2 nanoparticles, employing a variety of methods to enhance the photocatalytic and antimicrobial properties. The deposition of AgNPs can significantly improve the light-induced catalytic activity of TiO_2 nanoparticles. This could be ascribed to the ability of AgNPs to trap electrons at Schottky bar at each contact area of Ag/TiO_2 [16–19]. This results in a decrease in the recombination effect between electrons and holes on the surface of TiO_2 nanoparticles. Thus, separating the charge was stimulated and the transfer of electrons took place to result in a higher life-time of hole/electron pairs [20–22].

Viscose is a significant material for textiles owing to its high resistance to radiation and high stability to body fluids. The improvement of the antimicrobial properties of viscose has been critical for a variety of healthcare purposes. Therefore, various techniques have been reported to improve the antimicrobial properties of viscose fibers [23]. The weak binding of the colloidal nanoparticles to viscose fibers has been a substantial problem that can be overwhelmed by plasma treatment [24]. Plasma curing by etching was employed to activate the fibrous surfaces to induce the creation of polar groups, such as carbonyl, alcohol, carboxyl, and ether, facilitating better binding to nano-scaled particles [5,25]. Herein, we report the synthesis of TiO_2 NPs and Ag/TiO_2 nanocomposites as antibacterial and photocatalytic agents and their immobilization onto the surface of viscose fibers via a pad–dry–cure procedure to introduce multifunctional textiles. The morphologies and elemental compositions were evaluated by different analytical techniques. The performance of the Ag/TiO_2-coated viscose fibers showed an improved efficiency compared with TiO_2-coated viscose fibers.

2. Experimental details

2.1. Materials

Viscose fabrics were supplied from Spin and Weaving Misr El-Mahalla Co. (El-Mahalla City, Egypt) Silver(I) nitrate, titanium isopropoxide (TTIP; 97%), acetic acid (65%), silver nitrate (≥99.0%), acetic acid (CH_3COOH; 96%), nitric acid (HNO_3; 65%), sodium carbonate (Na_2CO_3), and oxalic acid were obtained from Aldrich (Cairo, Egypt). TiO_2 nanoparticles were synthesized according to the previously reported low temperature sol-gel method [18].

2.2. Synthesis of TiO_2 Nanoparticles

TTIP (2% v-v) was added to a solution of HNO_3 (1% v-v), CH_3COOH (10% v-v), and distilled water (DW). The mixture was subjected to stirring for an extra 16 h at 60 °C. After cooling, TiO_2 NPs were provided by continuously adding Na_2CO_3(aq) (5%) until reaching a full sedimentation of TiO_2 NPs. The generated dispersion was centrifuged (4000 rpm) for 5 min, decanted, washed with DW, and dried at 100 °C over 3 h. The dispersion of the generated TiO_2 NPs in DW was transparent and stable for several weeks at room temperature.

2.3. Synthesis of Ag/TiO_2

Ag/TiO_2 nanocomposite was synthesized utilizing UV technology [26]. Oxalic acid (0.005 mol/L) and $AgNO_3$ (0.0002 mol/L) in DW were mixed with a suspension of the above-prepared TiO_2 NPs (1 g). After stirring for 15 min, the mixture was added to

DW (450 mL) with vigorous stirring. The pH was adjusted in the range of 6.8–7.0 using NaOH(aq). The admixture was irradiated with UV supply for 60 min. The admixture was then placed to settle down for 8 h to form a brownish precipitation of Ag^0/TiO_2, which was filtered and dried at 120 °C for 3 h to give Ag/TiO_2.

2.4. Deposition of Ag/TiO$_2$ onto Plasma-Activated Viscose

As demonstrated in Figure 1, the plasma tool was applied to viscose fibers for 3 min at a power of 400 W and a constant pressure of 3×10^{-3} mbar [27]. The above-prepared solutions were then applied to the plasma-activated viscose fibers by the pad–dry–cure process. Both TiO_2 NPs (0.1 g) and Ag^0/TiO_2 (0.1 g) were stirred in DW (150 mL) and homogenized for 45 min under ambient conditions. The plasma-activated fabric (15 cm × 15 cm) was soaked in the prepared solutions for 60 min, and subjected to pad–dry–cure. The viscose was then dried at 90 °C, subjected to curing at 120 °C, and finally rinsed with DW. The binding stability of Ag^0/TiO_2 and TiO_2 onto viscose can be attributed to the electrostatic forces among Ti^{4+} existing on TiO_2 or Ag^0/TiO_2, and the negative charges on the viscose surface. The negative charges on viscose could be attributed to the negatively charged substituents, such as O–O– and –COO– generated by plasma.

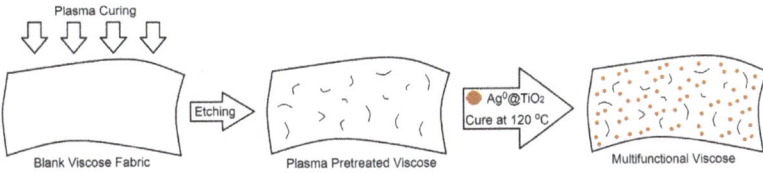

Figure 1. Schematic diagram representing the deposition of Ag^0/TiO_2 onto plasma-cured viscose fabric.

2.5. Characterization Methods

TEM (JEOL-1230, Akishima, Japan) was applied to inspect the morphology of the prepared TiO_2 NPs. The morphologies of the coated viscose were explored by Quanta SEM FEG 250 (Brno-Černovice, Czech Republic) linked to EDS (TEAM) to investigate the elemental contents of the viscose coated surface. FT-IR spectra were assessed by Nexus 670 (Nicolet; Watertown, MA, USA). UV/Vis absorption spectra and CIE Lab of the coated viscose were collected by UltraScanPro (Hunter Lab, Reston, VA, USA). The optical band gap was assessed from the absorbance spectrum utilizing Tauc's equation [$\varepsilon h\nu = C(h\nu - E_g)^n$], where E_g is the average band gap, ε is molar extinction coefficient, C is a constant, and n relies on the transition type.

2.6. Evaluation of Self-Cleaning

The self-cleaning activity was assessed by the light-induced decay of methylthioninium chloride (MTC) under visible (410 nm) and ultraviolet irradiation (315–380 nm) according to previous literature procedures [28].

2.7. Antibacterial Properties

The antibacterial performance was examined against *E. coli* according to the earlier procedure [29].

2.8. Durability Test

To study the durability of the treated viscose against washing, the coated samples (15 cm × 15 cm) were subjected to washing for 10 laundry cycles under AATCC 61:1989 standard procedure. The coated samples were charged in a laundry-o-meter machine, and subjected to washing with a detergent solution (200 mL) at 40 °C for 45 min. Both antibacterial and self-cleaning were assessed as indicators to evaluate the durability of the coated fibers.

3. Result and Discussion

3.1. Development of Composite

Ag/TiO$_2$ was synthesized under UV technology [19], starting from a mixture of oxalic acid, AgNO$_3$, and TiO$_2$ NPs (Figure 2). The reaction color was found to change from colorless to a brownish shade under irradiation with ultraviolet light to verify the reduction of silver ions (Ag$^+$) to silver metal (Ag^0) and incorporating AgNPs onto TiO$_2$ NPs. The color shift presented a visual verification for the photo-metallization process in the reaction system. Silver ions were initially subjected to cationic adsorption onto the surface of TiO$_2$ NPs. When a suspension of TiO$_2$ has a pH value <6, the main surface entities become TiOH^{2+}, while the main surface entities becomes TiOH$^-$ for a suspension of TiO$_2$ with a pH value higher than 6. Thus, NaOH(aq) was added for complete deposition of the adsorbed Ag$^+$ onto the surface of TiO$_2$ NPs to result in the formation of silver(I) oxide (Ag$_2$O), which were then reduced to Ag^0 by an ultraviolet supply. The ultraviolet irradiation has the ability to induce the transfer of free electrons from valence of TiO$_2$ NPs to the other conduction band. TiO$_2$ NPs comprises negative charges in the presence of Ti-OH$^-$, facilitating deposition of Ag$^+$ onto its surface. Thus, the photo-induced generated electrons function as reductive agents for Ag$^+$ to provide Ag^0. Production of tiny Ag^0 crystals could occur by cathode-like reduction or by aggregation of Ag^0. AgNPs has been known to show an absorbance band attributed to Plasmon effect owing to the interaction of the metallic NPs with UV, leading to oscillation of electrons. The color change of the solution to brown was attributed to the improved absorption at low wavelength owing to surface Plasmon. The reaction mechanism between AgNO$_3$ and TiO$_2$ is illustrated by the equations described below [30].

$$Ag^+ \rightarrow Ag^+ \text{ (adsorbed onto the surface of TiO}_2\text{ NPs)}$$
$$2Ag^+ \text{ (adsorbed)} + 2OH^- \rightarrow Ag_2O + H_2O$$
$$2Ag_2O + h\nu \rightarrow 4Ag^0 + O_2$$
$$TiO_2 \text{ (e-/h}^+\text{)} + Ag^+ \rightarrow TiO_2@Ag^0$$

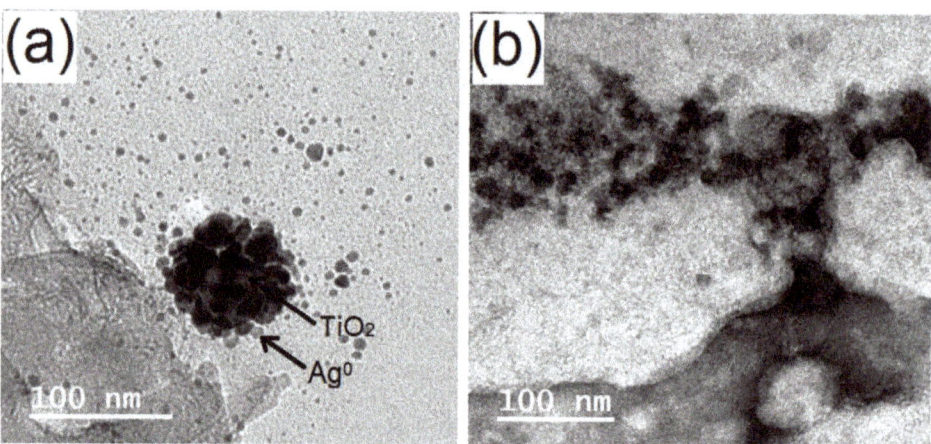

Figure 2. TEM graphs of TiO$_2$ (**a**) and Ag^0/TiO$_2$ (**b**).

UV/Vis absorption spectra were studied to explore the influence of Ag^0 on the TiO$_2$ optical activity, as illustrated in Figure 3. The absorption spectra of TiO$_2$ and Ag^0/TiO$_2$ showed broad absorbance bands with a wavelength maxima <400 nm. This can be attributed to the electron transition in TiO$_2$ depending on its energy band gap (~3.12 eV) owing to a charge transfer. The absorbance spectral curves of TiO$_2$ were enhanced in Ag^0/TiO$_2$. Obvious variations in absorbance activity of Ag^0/TiO$_2$ were detected in the visible spectrum range as a result of the weak Plasmon effect owing to the low Ag^0 content on TiO$_2$ NPs. This could enhance both surface excitation and electron/hole separation. The absorbance band of Ag^0/TiO$_2$ demonstrated that Ag^0/TiO$_2$ exhibits properties similar to TiO$_2$ NPs. The absorbance intensities were observed to exhibit a red shift for Ag^0/TiO$_2$, representing a decrease in TiO$_2$ gap. The absorbance spectra showed maximum absorbance wavelengths at 383 and 388 nm for TiO$_2$ and Ag^0/TiO$_2$, respectively. This monitored shift in the wavelength and the reduced band gap led to the increase in the light-induced catalytic activity of TiO$_2$ NPs in the visible range.

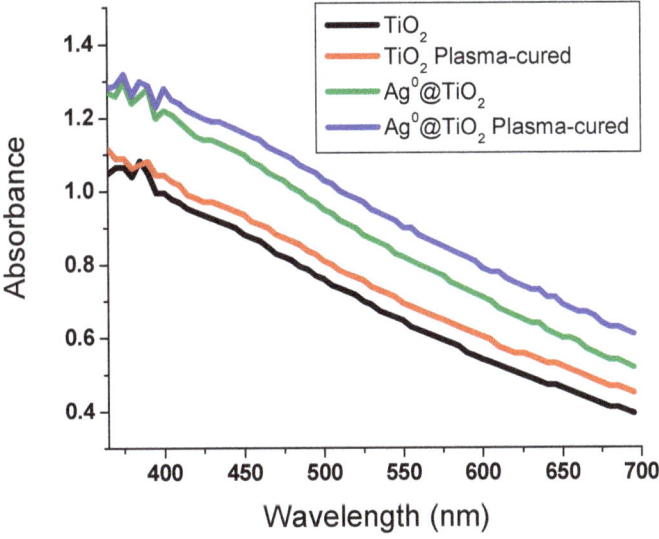

Figure 3. UV/Vis absorption spectral curves of the prepared composites coated onto viscose fibers.

3.2. Characterization of Viscose Fibers

The morphology of the coated viscose before and after treatment with plasma, as well as plasma-pretreated viscose before and after coating, were studied by SEM as depicted in Figure 4. A surface of moderate smoothness was monitored for plasma-inactivated viscose. Plasma-cured viscose displayed etches on the fiber surface. Irregular nanoparticles were monitored on the surface of the plasma-treated fibers. Decreasing the thickness of the surface layers resulted in improving the rough surface in comparison with pristine fibers. Fibers loaded with TiO$_2$ showed irregular and uneven clusters. Fibers coated with Ag^0/TiO$_2$ displayed a skinny film of inconsistent Ag^0/TiO$_2$. No cracking was detected and the small particles were monitored to cover the fibers. The changes in chemical compositions of samples due to plasma-curing and deposition of nanoparticles onto the surface of viscose were explored by EDX. The chemical compositions of blank and plasma-untreated fibers loaded with nanoparticles are summarized in Table 1. Both carbon and oxygen were detected as major contents due to the fabric, whereas Ti and/or Ag were detected as minor contents due to the deposition of TiO$_2$ or Ag^0/TiO$_2$ onto the fabric surface. The plasma-cured sample showed a slight increase in the oxygen content due to generating oxygen-containing substituents onto the fiber by oxygen plasma treatment. Plasma curing by etching and oxidation has been employed to activate the fiber surface to induce the creation of substituents [31], such as carbonyl, alcohol, carboxyl, and ether, facilitating strong binding to nano-scaled particles.

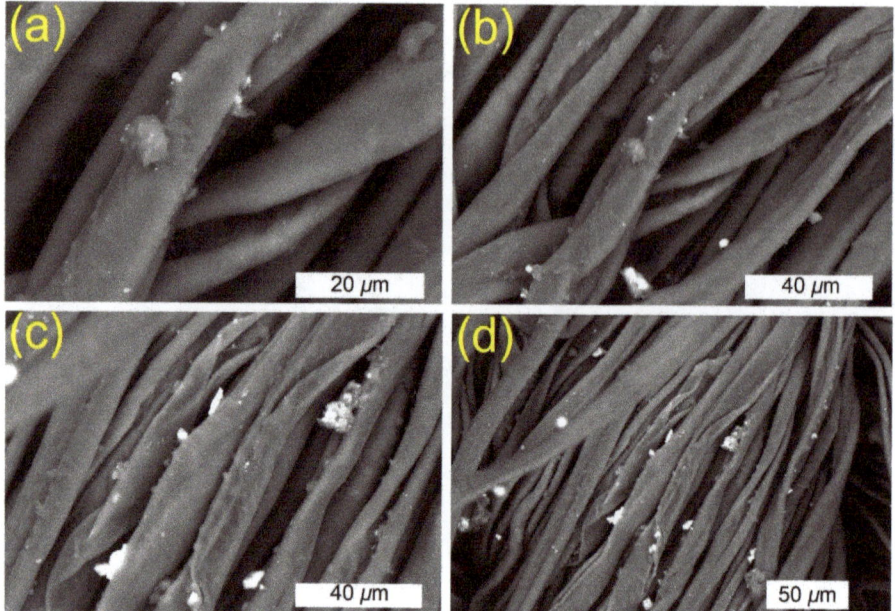

Figure 4. SEM images of TiO_2 NPs incorporated plasma-activated (**a,b**) and Ag^0/TiO_2 incorporated plasma-activated (**c,d**) fibers.

Table 1. Elemental contents of viscose fibers.

Sample	C	O	Ti	Ag
Blank	62.12 ± 1.3	37.88 ± 1.2	0	0
Plasma-activated	61.71 ± 1.1	38.29 ± 1.0	0	0
Plasma-inactivated (TiO_2)	59.44 ± 1.6	38.91 ± 1.1	1.65 ± 0.1	0
Plasma-activated (TiO_2)	57.14 ± 1.4	39.12 ± 1.6	3.74 ± 0.3	0
Plasma-inactivated (Ag^0/TiO_2)	59.03 ± 1.0	38.73 ± 1.3	1.72 ± 0.1	0.52 ± 0.1
Plasma-activated (Ag^0/TiO_2)	56.11 ± 1.2	39.43 ± 1.2	3.34 ± 0.2	1.12 ± 0.1

FT-IR spectra were explored for the coated viscose with and without plasma treatment, as shown in Figure 5. The main characteristic peaks were detected at 3339 cm^{-1} for the hydroxyl group stretch vibration, as well as two peaks at 2932 and 1030 cm^{-1} for the aliphatic C-H stretch and bend vibrations, respectively. No major shifts were detected in the absorbance bands; however, the intensity of the hydroxyl group was found to increase with the increasing deposition of the nanoparticles.

3.3. Self-Cleaning Properties

The reduction potential of MTC is about 0.011 V, whereas the energy level of the conduction band for TiO_2 is about −0.5 V. Thus, MTC is a suitable model to investigate the photo-induced catalysis process. The self-cleaning performance of TiO_2 or Ag^0/TiO_2 deposited onto viscose fibers could be studied by testing the decomposition of MTC underneath UV and visible lights, as shown in Figure 6. Ultraviolet/visible absorption spectral curves of MTC were collected for the treated viscose under irradiation with UV and visible light over 24 h. The absorbance peak at 665 nm decreased as a result of the degradation of MTC. The self-cleaning activity was tested by exploring the total content (C/C_0) of MTC as a function of time. The degradation of MTC on the uncoated fibers showed almost no variations under irradiation with either UV or visible daylight to prove that the uncoated fibers do not exhibit any light-induced decay ability. The deposition of TiO_2 NPs onto plasma-activated fibers proved an improvement in photo-induced degradation of MTC under UV light. However, this photo-induced degradation of MTC was incomplete. The deposition of TiO_2 onto plasma-pretreated fibers displayed a negligible photo-induced degradation under visible light owing to the adsorption and diffusion of MTC within the coated viscose. The integration of the nanocomposite into the plasma-pretreated fibers induced a total photoinduced degradation of MTC under ultraviolet and visible light, as the blue shade was monitored neither on the coated fibers nor in solution to prove a complete photo-induced degradation of MTC. The photo-induced degradation of MTC for fibers coated with Ag^0/TiO_2 demonstrated improved activity in comparison with TiO_2 NPs, proposing that the inclusion of AgNPs onto the surface of TiO_2 is an efficient approach. The photo-induced degradation rate for TiO_2 and Ag^0/TiO_2 coated onto viscose fibers decreased with washing. Nonetheless, they persisted higher than the plasma-inactivated viscose fibers. This proposed higher adhesion of particles onto plasma-activated viscose. After washing, the light-induced decay of methylthioninium chloride (MTC) for Ag^0/TiO_2 deposited onto viscose fibers was lower than the case of the viscose fibers coated with TiO_2 under visible/UV light. Thus, the nanocomposite enhanced the self-cleaning activity of viscose as a beneficial effect of silver on the light-induced catalysis of TiO_2.

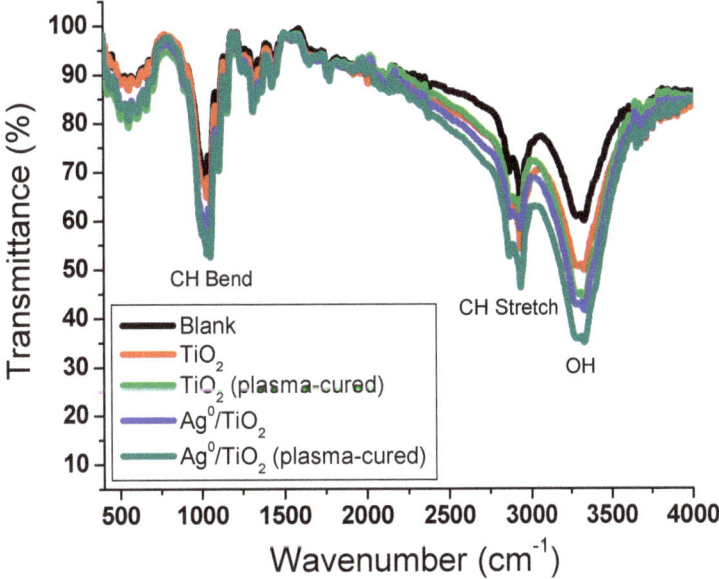

Figure 5. FT-IR spectra of coated viscose fibers.

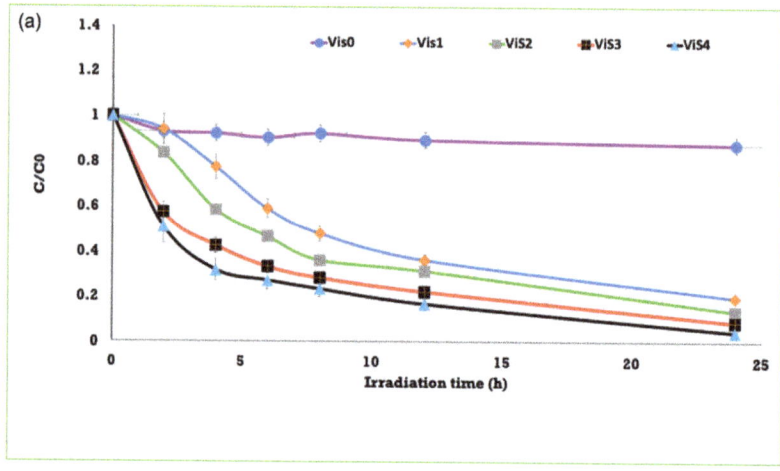

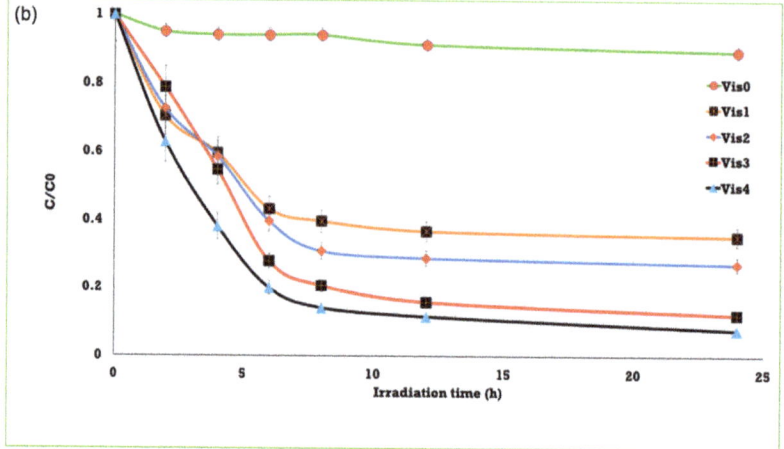

Figure 6. Degradation of MTC on fibers under UV (**a**) and visible (**b**) lights for pristine fibers (Vis$_0$), TiO$_2$/fibers after wash (Vis$_1$), TiO$_2$/fibers before wash (Vis$_2$), Ag^0/TiO$_2$/fibers after wash (Vis$_3$), and Ag^0/TiO$_2$/fibers before wash (Vis$_4$).

3.4. Antibacterial Activity

The antibacterial properties of plasma-cured and coated viscose were examined against *E. coli* by measuring optical density (OD) at 620 nm versus time, as shown in Figure 7. OD was found to improve, reflecting the decrease in the quantity of growing bacteria in the tested sample. Both blank and TiO$_2$ coated viscose fibers displayed no inhibition. The viscose fibers coated with Ag^0/TiO$_2$ showed antibacterial properties at all contents, yet followed by washing to confirm the positive effect of loading Ag^0 onto TiO$_2$. AgNPs have been described to exhibit a broad of activity against a variety of pathogens. It has been recognized that the increase in surface area results in improved antibacterial properties [32].

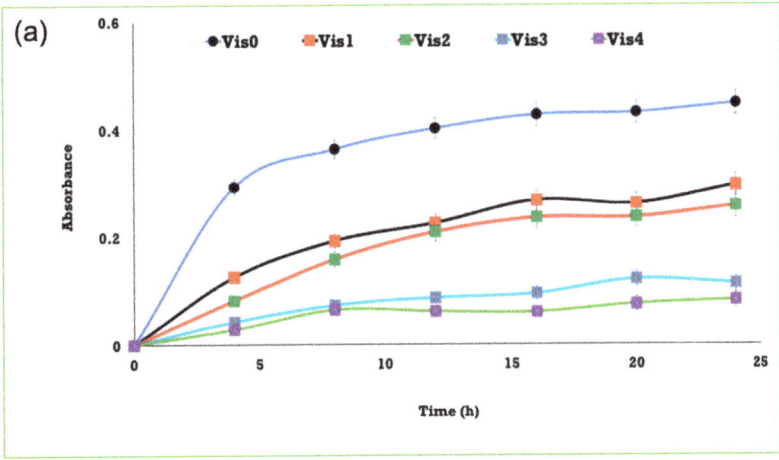

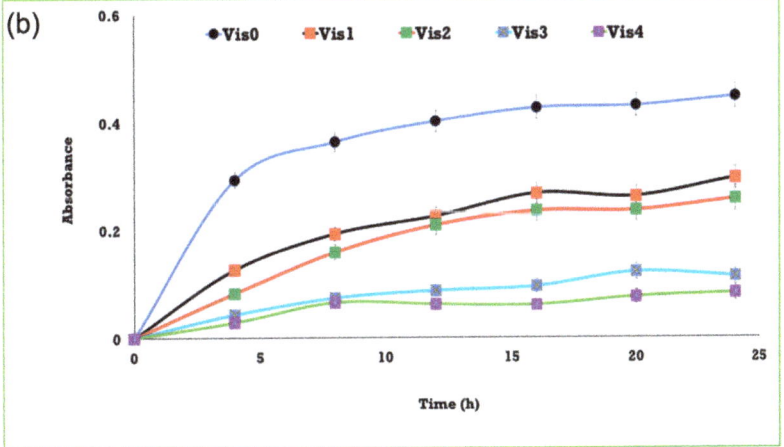

Figure 7. Activity of coated viscose fibers against *E. coli*; Vis$_0$ is pristine viscose, Vis$_1$ is 10^3 (**a**) or 10^6 (**b**) bacterial density, Vis$_2$ is NPs/fibers, Vis$_3$ is nanocomposite/fibers following washing, and Vis$_4$ is nanocomposite/fibers prior to washing.

4. Conclusions

Multifunctional viscose fibers coated with Ag/TiO$_2$ nanocomposite were developed by the simple pad–dry–cure technology. The synthesis, characterization, and use of nanocomposite as an antibacterial and light-induced self-cleaning agent were explored. Ag/TiO$_2$ was prepared using a double-stage procedure of sol–gel TiO$_2$ synthesis, followed by depositing of Ag^0 onto the surface of TiO$_2$ by ultraviolet irradiation. The deposition of Ag^0/TiO$_2$ onto plasma-pretreated viscose fibers was accomplished using the facile pad–dry cure technology. Ag^0/TiO$_2$ displayed better absorption in the visible spectrum and higher antibacterial activity and light-induced catalysis in comparison with plasma-activated viscose coated with TiO$_2$. This considerable improvement in antibacterial and self-clean properties could be attributed to AgNPs deposited onto the surface of TiO$_2$. The current study presented a good strategy to produce Ag/TiO$_2$ composite with the ability to impart antibacterial, self-cleaning photo-induced catalytic properties to plasma-cured fibers, under irradiation with UV/visible lights to make this Ag^0/TiO$_2$ nanocomposite potentially practical as a multifunctional agent for a variety of applications, such as medical clothing.

Author Contributions: Conceptualization H.E.-H. and M.E.E.-N.; methodology, M.E.E.-N., T.A.K. and A.E.-F.; formal analysis and discussion, M.E.E.-N., A.E.-F. and T.A.K.; writing—original draft preparation, H.E.-H. and M.E.E.-N.; writing—review and editing, H.E.-H., M.E.E.-N., T.A.K. and A.E.-F.; supervision, H.E.-H. and M.E.E.-N. All authors have read and agreed to the published version of the manuscript.

Funding: Deanship of Scientific Research at King Saud University, Research Group Program (no. RGP-201), King Saud University, Riyadh, Saudi Arabia.

Institutional Review Board Statement: Not applicable.

Informed Consent Statement: Not applicable.

Data Availability Statement: The data presented in this study are available on request from the corresponding author.

Acknowledgments: The authors would like to extend their appreciation to the Deanship of Scientific Research at King Saud University for funding this work through research group (no. RGP-201).

Conflicts of Interest: The authors declare no conflict of interest.

References

1. Yasin, S.; Sun, D. Propelling textile waste to ascend the ladder of sustainability: EOL study on probing environmental parity in technical textiles. *J. Clean. Prod.* **2019**, *233*, 1451–1464. [CrossRef]
2. Sharaf, S.; El-Naggar, M.E. Wound dressing properties of cationized cotton fabric treated with carrageenan/cyclodextrin hydrogel loaded with honey bee propolis extract. *Int. J. Biol. Macromol.* **2019**, *133*, 583–591. [CrossRef]
3. Hebeish, A.; El-Naggar, M.E.; Tawfik, S.; Zaghloul, S.; Sharaf, S. Hyperbranched polymer–silver nanohybrid induce super antibacterial activity and high performance to cotton fabric. *Cellulose* **2019**, *26*, 3543–3555. [CrossRef]
4. El-Naggar, M.E.; Abdelgawad, A.M.; Elsherbiny, D.A.; El-shazly, W.A.; Ghazanfari, S.; Abdel-Aziz, M.S.; Abd-Elmoneam, Y.K. Bioactive Wound Dressing Gauze Loaded with Silver Nanoparticles Mediated by Acacia Gum. *J. Clust. Sci.* **2019**. [CrossRef]
5. El-Naggar, M.E.; Khattab, T.A.; Abdelrahman, M.S.; Aldalbahi, A.; Hatshan, M.R. Development of antimicrobial, UV blocked and photocatalytic self-cleanable cotton fibers decorated with silver nanoparticles using silver carbamate and plasma activation. *Cellulose* **2021**, *28*, 1105–1121. [CrossRef]
6. Chen, G.; Li, Y.; Bick, M.; Chen, J. Smart textiles for electricity generation. *Chem. Rev.* **2020**, *120*, 3668–3720. [CrossRef]
7. Lee, D.; Sang, J.S.; Yoo, P.J.; Shin, T.J.; Oh, K.W.; Park, J. Machine-washable smart textiles with photothermal and antibacterial activities from nanocomposite fibers of conjugated polymer nanoparticles and polyacrylonitrile. *Polymers* **2019**, *11*, 16. [CrossRef] [PubMed]
8. Hiremath, L.; Kumar, S.N.; Sukanya, P. Development of antimicrobial smart textiles fabricated with magnetite nano particles obtained through green synthesis. *Mater. Today Proc.* **2018**, *5*, 21030–21039. [CrossRef]
9. Moazzenchi, B.; Montazer, M. Click electroless plating and sonoplating of polyester with copper nanoparticles producing conductive fabric. *Fibers Polym.* **2020**, *21*, 522–531. [CrossRef]
10. Dhineshbabu, N.R.; Bose, S. Smart textiles coated with eco-friendly UV-blocking nanoparticles derived from natural resources. *ACS Omega* **2018**, *3*, 7454–7465. [CrossRef] [PubMed]
11. Ahmed, H.; Khattab, T.A.; Mashaly, H.M.; El-Halwagy, A.A.; Rehan, M. Plasma activation toward multi-stimuli responsive cotton fabric via in situ development of polyaniline derivatives and silver nanoparticles. *Cellulose* **2020**, *27*, 2913–2926. [CrossRef]
12. Elwakeel, K.Z.; Abd El-Ghaffar, M.A.; El-Kousy, S.M.; El-Shorbagy, H.G. Enhanced remediation of Reactive Black 5 from aqueous media using new chitosan ion exchangers. *J. Dispers. Sci. Technol.* **2013**, *34*, 1008–1019. [CrossRef]
13. Rabie, S.T.; Ahmed, A.E.; Sabaa, M.W.; Abd El-Ghaffar, M.A. Maleic diamides as photostabilizers for polystyrene. *J. Ind. Eng. Chem.* **2013**, *19*, 1869–1878.
14. Chen, H.; Huang, M.; Wang, Z.; Gao, P.; Cai, T.; Song, J.; Zhang, Y.; Meng, L. Enhancing rejection performance of tetracycline resistance genes by a TiO_2/AgNPs-modified nanofiber forward osmosis membrane. *Chem. Eng. J.* **2020**, *382*, 123052. [CrossRef]
15. Jatoi, A.W.; Kim, I.S.; Ni, Q.-Q. Cellulose acetate nanofibers embedded with AgNPs anchored TiO_2 nanoparticles for long term excellent antibacterial applications. *Carbohydr. Polym.* **2019**, *207*, 640–649. [CrossRef] [PubMed]
16. Nguyen, V.T.; Vu, V.T.; Nguyen, T.A.; Tran, V.K.; Nguyen-Tri, P. Antibacterial activity of TiO_2-and ZnO-decorated with silver nanoparticles. *J. Compos. Sci.* **2019**, *3*, 61. [CrossRef]
17. Djellabi, R.; Basilico, N.; Delbue, S.; D'Alessandro, S.; Parapini, S.; Cerrato, G.; Laurenti, E.; Falletta, E.; Bianchi, C.L. Oxidative Inactivation of SARS-CoV-2 on Photoactive AgNPs@ TiO_2 Ceramic Tiles. *Int. J. Mol. Sci.* **2021**, *22*, 8836. [CrossRef] [PubMed]
18. Chou, J.-C.; Lin, Y.-C.; Lai, C.-H.; Kuo, P.-Y.; Nien, Y.-H.; Syu, R.-H.; Yong, Z.-R.; Wu, Y.-T. AgNWs@ TiO_2 and AgNPs@ TiO_2 Double-Layer Photoanode Film Improving Light Capture and Application under Low Illumination. *Chemosensors* **2021**, *9*, 36. [CrossRef]
19. El-Naggar, M.E.; Hasanin, M.; Youssef, A.M.; Aldalbahi, A.; El-Newehy, M.H.; Abdelhameed, R.M. Hydroxyethyl cellulose/bacterial cellulose cryogel dopped silver@titanium oxide nanoparticles: Antimicrobial activity and controlled release of Tebuconazole fungicide. *Int. J. Biol. Macromol.* **2020**, *165*, 1010–1021. [CrossRef] [PubMed]

20. Kusiak-Nejman, E.; Czyżewski, A.; Wanag, A.; Dubicki, M.; Sadłowski, M.; Wróbel, R.J.; Morawski, A.W. Photocatalytic oxidation of nitric oxide over AgNPs/TiO2-loaded carbon fiber cloths. *J. Environ. Manag.* **2020**, *262*, 110343. [CrossRef] [PubMed]
21. Abdelwahab, N.A.; El-Nashar, D.E.; El-Ghaffar, M.A.A. Polyfuran, polythiophene and their blend as novel antioxidants for styrene-butadiene rubber vulcanizates. *Mater. Des.* **2011**, *32*, 238–245. [CrossRef]
22. El-Enany, G.M.; Ghanem, M.A.; El-Ghaffar, M.A. Electrochemical deposition and characterization of poly (3,4-ethylene dioxythiophene), poly (aniline) and their copolymer onto glassy carbon electrodes for potential use in ascorbic acid oxidation. *Port. Electrochim. Acta* **2010**, *28*, 336–348.
23. Li, W.; Yu, Z.; Wu, Y.; Liu, Q. Preparation, characterization of feather protein-g-poly (sodium allyl sulfonate) and its application as a low-temperature adhesive to cotton and viscose fibers for warp sizing. *Eur. Polym. J.* **2020**, *136*, 109945. [CrossRef]
24. Revol, B.P.; Vauthier, M.; Thomassey, M.; Bouquey, M.; Ruch, F.; Nardin, M. Design of experience to evaluate the Interfacial compatibility on high tenacity viscose fibers reinforced Polyamide-6 composites. *Compos. Sci. Technol.* **2021**, *203*, 108615. [CrossRef]
25. Katouah, H.; El-Metwaly, N.M. Plasma treatment toward electrically conductive and superhydrophobic cotton fibers by in situ preparation of polypyrrole and silver nanoparticles. *React. Funct. Polym.* **2021**, *159*, 104810. [CrossRef]
26. Hernández-Gordillo, A.; Arroyo, M.; Zanella, R.; Rodríguez-González, V. Photoconversion of 4-nitrophenol in the presence of hydrazine with AgNPs-TiO2 nanoparticles prepared by the sol–gel method. *J. Hazard. Mater.* **2014**, *268*, 84–91. [CrossRef] [PubMed]
27. Atta, A.M. Immobilization of silver and strontium oxide aluminate nanoparticles integrated into plasma-activated cotton fabric: Luminescence, superhydrophobicity, and antimicrobial activity. *Luminescence* **2021**, *36*, 1078–1088. [CrossRef] [PubMed]
28. Wu, S.; Huang, J.; Cui, H.; Ye, T.; Hao, F.; Xiong, W.; Liu, P.; Luo, H. Preparation of organic–inorganic hybrid methylene blue polymerized organosilane/sepiolite pigments with superhydrophobic and self-cleaning properties. *Text. Res. J.* **2019**, *89*, 4220–4229. [CrossRef]
29. Atta, A.M.; Abomelka, H.M. Multifunctional finishing of cotton fibers using silver nanoparticles via microwave-assisted reduction of silver alkylcarbamate. *Mater. Chem. Phys.* **2021**, *260*, 124137. [CrossRef]
30. Katouah, H.A.; El-Sayed, R.; El-Metwaly, N.M. Solution blowing spinning technology and plasma-assisted oxidation-reduction process toward green development of electrically conductive cellulose nanofibers. *Environ. Sci. Pollut. Res.* **2021**, 1–13. [CrossRef]
31. Chiam, S.-L.; Soo, Q.-Y.; Pung, S.-Y.; Ahmadipour, M. Polycrystalline TiO$_2$ particles synthesized via one-step rapid heating method as electrons transfer intermediate for Rhodamine B removal. *Mater. Chem. Phys.* **2021**, *257*, 123784. [CrossRef]
32. Biniaś, D.; Biniaś, W.; Machnicka, A.; Hanus, M. Preparation of antimicrobial fibres from the EVOH/EPC blend containing silver nanoparticles. *Polymers* **2020**, *12*, 1827. [CrossRef] [PubMed]

Article

Multifunctional Dyeing of Wool Fabrics Using Selenium Nanoparticles

Tarek Abou Elmaaty [1,2], Sally Raouf [3], Khaled Sayed-Ahmed [4] and Maria Rosaria Plutino [5,*]

[1] Department of Material Arts, Faculty of Art & Design, Galala University, Galala 43713, Egypt; tasaid@gu.edu.eg
[2] Department of Textile Printing, Dyeing & Finishing, Faculty of Applied Art, Damietta University, Damietta 34511, Egypt
[3] Department of Textile Printing, Dyeing & Finishing, Faculty of Applied Art, Banha University, Banha 13518, Egypt; soola.a@hotmail.com
[4] Department of Agricultural Chemistry, Faculty of Agriculture, Damietta University, Damietta 34511, Egypt; dr_khaled@yahoo.com
[5] Consiglio Nazionale Delle Ricerche, c/o Dipartment ChiBioFarAm, Istituto per lo Studio dei Materiali Nanostrutturati, University of Messina, Viale F. D'Alcontres 31, Vill. S. Agata, 98166 Messina, Italy
* Correspondence: mariarosaria.plutino@cnr.it; Tel.: +39-(09)-06765713

Abstract: This work aims to utilize selenium nanoparticles (Se-NPs) as a novel dyestuff, which endows wool fibers with an orange color because of their localized surface plasmon resonance. The color characteristics of dyed fibers were evaluated and analyzed. The color depth of the dyed fabrics under study was increased with the increase in Se content and dyeing temperature. The colored wool fabrics were characterized using scanning electron microscopy (SEM), energy dispersive spectroscopy (EDX) and an X-ray diffraction (XRD) analysis. The results indicated that spherical Se-NPs with a spherical shape were consistently deposited onto the surface of wool fibers with good distribution. In addition, the influence of high temperature on the color characteristics and imparted functionalities of the dyed fabrics were also investigated. The obtained results showed that the proposed dyeing process is highly durable to washing after 10 cycles of washes, and the acquired functionalities, mainly antimicrobial activity and UV-blocking properties, were only marginally affected, maintaining an excellent fastness property.

Keywords: multifunctional finishing; selenium nanoparticles; wool fabrics; UV-blocking properties; antimicrobial activity

Citation: Elmaaty, T.A.; Raouf, S.; Sayed-Ahmed, K.; Plutino, M.R. Multifunctional Dyeing of Wool Fabrics Using Selenium Nanoparticles. *Polymers* 2022, *14*, 191. https://doi.org/10.3390/polym14010191

Academic Editor: Sándor Kéki

Received: 30 November 2021
Accepted: 31 December 2021
Published: 4 January 2022

Publisher's Note: MDPI stays neutral with regard to jurisdictional claims in published maps and institutional affiliations.

Copyright: © 2022 by the authors. Licensee MDPI, Basel, Switzerland. This article is an open access article distributed under the terms and conditions of the Creative Commons Attribution (CC BY) license (https://creativecommons.org/licenses/by/4.0/).

1. Introduction

Recently, the usage of unique dyes and finishing agents has had a substantial influence on textiles' functionality. The dyeing of textiles with multi-functional characteristics has attracted significant interest in recent years. Nanotechnologies and nanomaterials offer a wider application potential for preparing functional textiles, such as flame retardant, self-cleaning, wrinkle-resistant, antistatic, antimicrobial, and UV-protective, etc. [1–14] In recent years, several approaches to metal and metal oxide NP fabrication and application have been developed to impart functionalities to different types of textile substrates [15–23]. In addition, nanomaterials, such as gold and silver nanoparticles, stabilized by opportune functional capping agents, have been used as a novel class of functionalizing nanofillers either themselves [24–29] or for the functionalization of different types of materials (such as concrete, glass, geopolymer and also embedded in coloring or antibacterial coatings), especially for textile fabrics [30].

The coloration of fabric/fiber can be achieved through a combination with metal nanoparticles by means of the particular optical properties of plasmon nanomaterials, called localized surface plasmon resonance (LSPR), producing brilliant and vivid colors, which are different from traditional dyes in that they are not the chromophore of traditional dyes but the shape and size of nanoparticles that determine the colors [31,32]. This has recently motivated research activities to directly employ metal nanoparticles in the dyeing of different textile substrates to overcome color fading and high water and energy consumption problems associated with conventional dyeing methods. Furthermore, excessive exposure to dyes may cause respiratory problems and skin irritation and increase risk of cancer [33]. Hence, several different approaches have been developed to impart colors and functionalities to diverse fabrics/fibers, such as cotton, wool [34–41], silk [42–45], bamboo [46], ramie [47], viscose and acrylic [48] using noble metal nanoparticles via a self-assembly or in-situ synthesis methods.

Owing to the localized Surface Plasmon Resonance (LSPR) property of selenium nanoparticles (Se-NPs), they exhibit adaptable colors depending on the synthesizing measures. Besides, Se-NPs have strong cytotoxicity towards a broad range of microorganisms, low toxicity to human cells, high selectivity, and long-term durability [49]. Se-NPs have significant antimicrobial and antioxidant activity, which have been widely reported in the literature [50–52], and their applications in textile fabrics have been reported once by Joanne Yip et al. [53] as an antimicrobial agent for polyester fabric.

In our previous study, cotton and polyester were colored with silver and gold NPs via a printing technique to render different functions of fibrous materials [54]. Subsequently, Se-NPs were used in the wool fabric printing process to fabricate antimicrobial textiles with low cytotoxicity [55]. In the current study, wool fabrics were dyed with Se-NPs in a different way to produce multifunctional fabrics with brilliant stable colors in one step and compare between the results obtained from both dyeing and printing techniques.

2. Experimental

2.1. Materials

The fabrics used here were 100% scoured and bleached wool. Sodium hydrogen selenite, vitamin C and polyvinylpyrrolidone (PVP) were purchased from *Loba Chemie, India*. In addition, the other chemicals used in this study were of commercial grade.

2.2. Dyeing Apparatus

Dyeing was carried out using an Infra Color Dyeing Machine, which consists of 12 beakers that are mounted in a rotating beaker-carrying wheel. Heating occurs through infrared radiation, cooling through air and automation through the microprocessor programmer DC4 F/R. The maximum temperature was up to 140 °C, a maximum heating rate was up to 5 °C/min and cooling had a maximum rate up to 3 °C/min.

2.3. Methods

2.3.1. Green Synthesis of Se-NPs

Se-NPs were prepared according to the method reported by Abou Elmaaty et al. [55] through a redox reaction. Sodium hydrogen selenite at different concentrations of 25 mM, 50 mM, 75 mM and 100 mM was added to vitamin C at the same concentration and the volume ratio of 1:1 under magnetic stirring. PVP was used as a stabilizer at the concentration of 0.75–3 g/100mL of vitamin C to enhance the stability of Se-NPs. The change in color from colorless to orange to dark orange indicated the formation of Se-NPs.

2.3.2. Dyeing Procedure of Wool Fabrics with Selenium NPs

Wool fabrics were dyed with different concentrations of Se-NP solution using a wet chemistry method through an immersion process. Briefly, the wool fabrics were immersed into the Se-NP solutions (dyeing solutions) without any chemical additives. Dyeing was performed in a laboratory-scale thermal HT dyeing machine with a liquor-to-goods ratio of 100:1. The solutions containing Se-NPs and wool fabrics were shaken at different temperatures, including room temperature, 40 °C, 70 °C and 100 °C for 60 min to obtain the coloration of fabrics. At the end of the dyeing process, the dyed wool samples were removed, rinsed with water and dried at room temperature.

2.4. Characterization

TEM images of the synthesized Se-NPs were obtained using a JEM-2100 Transmission Electron Microscope (TEM) with an acceleration voltage of 200 kV to determine the size and morphology of the NPs. Samples for the TEM analysis were prepared by dripping a drop of Se-NP solution onto a carbon-coated copper grid and drying at room temperature.

The surface morphology of wool fabrics, either blank or dyed fabrics, was examined using high resolution scanning electron microscopy (JEOL JSM-6510LB with a field emission gun, Tokyo, Japan). The chemical structure of the dyed samples was analyzed using a surface energy dispersive spectroscopy (EDX) analysis unit (EDAX AMETEK analyzer, Rigaku, Japan) attached to the electron microscope.

X-ray diffraction was analyzed for both the prepared Se-NPs and Se-dyed wool fabrics using an X-ray diffractometer system (XRD) (Bruker D8 ADVANCE, Karlsruhe, Germany). The Se-NP solution was dried at 130 °C until completely dried.

2.5. Functional Properties of Se-Dyed Wool Fabrics

The functionalities of wool fabrics dyed with Se-NPs were evaluated in terms of the UV protection factor (UPF) and antimicrobial activity.

2.5.1. Antimicrobial Activity

The biological activity of the Se-dyed wool samples was evaluated qualitatively against G+ve bacteria (Staphylococcus aureus and Bacillus cereus), G-ve bacteria (Escherichia coli) and yeast (Candida utilis). The antimicrobial test was performed according to the AATCC Test Method (147–1988). The antimicrobial activity was expressed as the growth inhibition zone (mm).

2.5.2. UV-Protection Properties

The ultraviolet protection factor expressed as UPF and transmission of ultraviolet (UV) radiation through wool fabrics were evaluated based on the AS/NZS 4399:1996 test method. The UV protection was rated as good, very good or excellent if the UPF values were between 15–24, 25–39 or above 40, respectively.

2.6. Testing

2.6.1. Color Measurements

The color coordinates of CIE lab (L^*, a^*, b^*, C^*, and h^*) and color strength (K/S) for both blank and Se-dyed wool fabrics were measured in the wavelength range of 360–720 nm using a Konica Minolta spectrophotometer (CM-3600 d, Minolta, Tokyo, Japan). All samples were measured in three different areas, bearing in mind both sides of the fabrics, and the mean values were recorded.

2.6.2. Fastness Properties

The color fastness properties were evaluated for all Se-dyed wool fabrics in accordance with the standard test method. Washing, rubbing and light fastness were measured according to the method of AATCC test methods (61–1972), (8–1972) and (16A–1972), respectively.

2.6.3. Durability Test

Durability to washing of the imparted functionalities (antimicrobial activity and UV-protection properties) as well as color strength (K/S) of Se-dyed wool fabrics were evaluated according to AATCC Test Method 61(2A)-1996 after 10 laundering cycles.

2.6.4. Mechanical Properties

The tensile strength (maximum load) and elongation at break (maximum strain) of blank and Se-dyed wool fabrics at high temperature were evaluated using the strip test method on a multi-tester machine according to ASTM-D4850 with a load cell 500 N, preload 0.01 N, speed of 100 mm/min and gauge length of 100 mm.

2.7. Statistical Analysis

Antimicrobial, UV-protection, and mechanical tests were performed by taking the average of three readings (samples). The standard error of the mean was calculated according to the equation given below and was found to be + (−) 0.1 SEX = S/$\sqrt{n}$

Where S = sample standard deviation and n = the number of observations of the samples.

3. Results and Discussion

The dyeing process of wool fabrics using selenium nanoparticles (Se-NPs) as a new functional colorant was investigated to obtain multifunctional dyed fabrics based on colloidal solutions of nanomaterials. The influences of different dyeing temperatures on the adsorption of Se-NPs onto wool fabrics and alteration of the color characteristics of wool samples were observed. The imparted functionalities were also determined. The results obtained and appropriate discussions are presented below.

3.1. Characterization of Se-NPs and Dyed Wool Fabrics

3.1.1. Transmission Electron Microscopy (TEM)

Figure 1 shows TEM images of colloidal Se-NPs prepared at different concentrations. The obtained TEM micrographs revealed that the Se-NPs prepared at a concentration of 50 mM had the lowest diameter range (25–90 nm) compared to the other prepared Se-NPs.

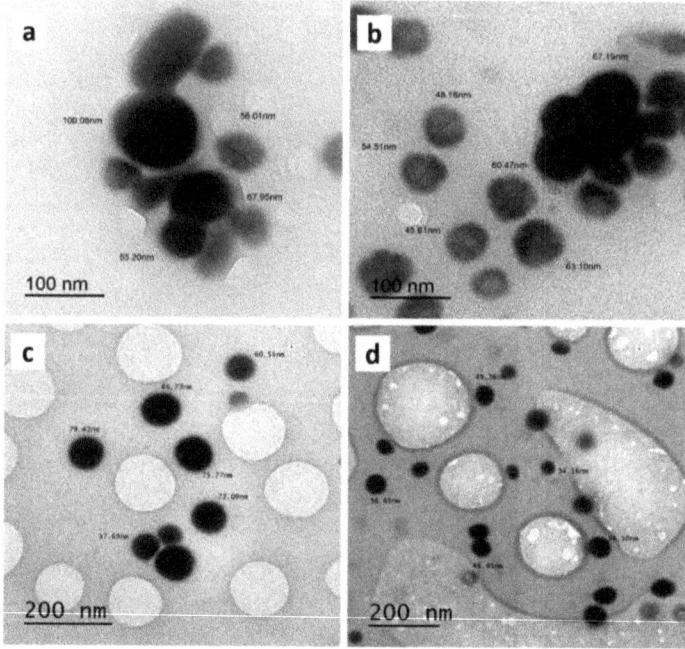

Figure 1. TEM images of synthesized Se-NPs on different concentrations: (**a**) 12.5 mM, (**b**) 25 mM, (**c**) 37.5 mM and (**d**) 50 mM.

The Se-NPs were well-dispersed and mostly spherical in shape when no agglomeration or deformation of Se-NPs were observed. The Se-NPs prepared at a concentration of 12.5 mM showed the largest size up to 115 nm, resulting in the appearance of a block structure, as shown in Figure 1a. While the concentration of the stabilizer PVP (3 g/100mL) and sodium hydrogen selenite increased, the shape of the Se-NPs was observed to be uniformly spherical instead of an aggregated form and had the lowest diameter, as shown in Figure 1d.

3.1.2. SEM and EDX Analysis

In Figure 2, SEM images show the topographical characteristics of the blank and Se-dyed wool fabrics. The SEM images of the blank wool show typically clear, clean scales and a smooth longitudinal fibrous structure surface, as shown in Figure 2(0). The SEM images of the surface of the Se-dyed wool fabrics revealed that the surfaces of wool fabrics were covered by a sufficient layer of Se in the nano size, and Se-NPs on the wool fabric surfaces had a wide range of size distribution.

The surface chemical elements of the Se-dyed wool fabrics were determined by EDX spectroscopy, as shown in Figure 2e–h. The peaks appearing at about 1 and 11 Kev in each figure are attributed to selenium NPs. The SEM images and EDX results provide strong evidence for the formation of Se-NPs on the surface of the wool fabrics.

3.1.3. X-ray Diffraction (XRD)

An XRD analysis was performed for further confirmation of the Se-NP formation. This analytical method aids in the determination of crystallite materials and provides details of the unit cell dimensions. As shown in Figure 3, the obtained Se-NPs for both prepared solutions and Se-dyed wool fabrics were highly crystalline, and all diffraction peaks were well indexed as $24.28°$, $29.24°$ and $43.64°$, corresponding to 100, 101 and 102 crystal planes, respectively, in accordance with the JCPDS 86-2246 international database. [56] The results obtained from the SEM, EDX and XRD analyses confirmed the sufficient deposition of Se-NPs on the wool fabric surface.

3.2. Functional Properties of Se-NP Dyed Wool Fabrics

3.2.1. Biological Activity

The large surface area of natural textile fabrics and their ability to retain moisture provide an excellent environment for microorganism growth. Therefore, imparting the antimicrobial properties for natural fabrics is of great interest. Thus, the antimicrobial activity of the unwashed and washed Se-dyed wool fabrics was evaluated qualitatively against G+ve (*S. aureus* and & *Bacillus cereus*), G-ve (*E. coli*) and yeast (*Candida utilis*). The results are reported in Table 1 and are expressed as the zone of growth inhibition ZI (mm).

From the results, it is quite clear that the blank wool fabric did not show any antimicrobial effect. However, the Se-dyed wool showed a notable antimicrobial activity that increased with the increase in Se-NP concentration against the tested pathogens. In this respect, SEM micrographs illustrated that the deposition of Se-NPs on the wool surface increased with the increase in Se-NP concentration, confirming the results obtained from the antimicrobial test. This enhancement may be due to the adsorption of Se-NPs, leading to cell wall depolarization, which changes the typically negative charge of the wall to become more permeable, and then, inhibition of cell membrane metabolisms causing the death of bacteria, and/or the increase in NP concentrations leading to a concomitant increase of highly reactive oxygen radicals causing the destruction of the molecular structure of bacteria [57].

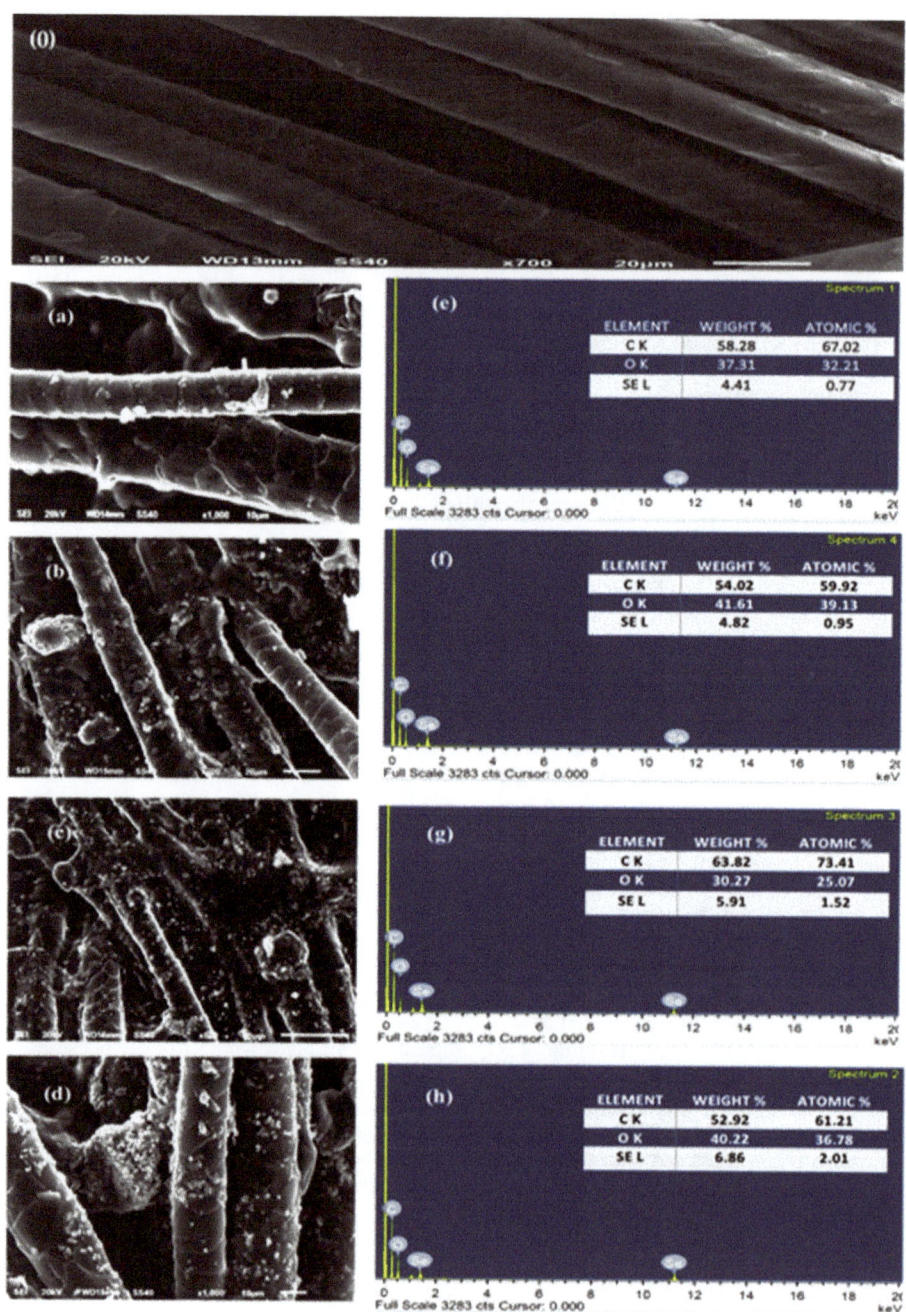

Figure 2. SEM micrographs of (**0**) blank wool fabric, and (**a–d**) wool fabrics dyed with Se-NPs at different concentrations of: (**a**) 12.5 mM, (**b**) 25 mM, (**c**) 37.5 mM and (**d**) 50 mM, as well as EDX spectra of Se-dyed wool fabrics with varied concentrations of: (**e**) 12.5 mM, (**f**) 25 mM, (**g**) 37.5 mM and (**h**) 50 mM.

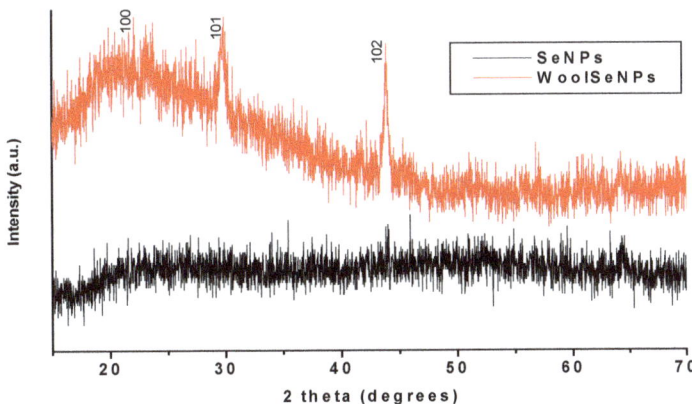

Figure 3. XRD patterns of the synthesized Se-NPs and Se-dyed wool fabric.

Table 1. Antimicrobial activity of the Se-dyed wool fabrics at different temperatures.

Se-NPs Conc.	Temp. °C	Antimicrobial Activity ZI [1] (mm.)			
		G+ve		G−ve	Yeast
Blank wool		S. aureus	Bacillus cereus	E. coli	Candida utilis
		0	0	0	0
12.5 mM		13.4 (13) [2]	10.6 (9)	10 (10)	14 (13.5)
25 mM	At room	15.9 (15)	12.6 (12)	13.5 (12.4)	17.5 (16)
37.5 mM	temp.	17.7 (17)	14 (13.2)	14.3 (14)	19 (18)
50 mM		19 (18)	15.4 (14.8)	18.2 (18)	23 (22.3)
12.5 mM		13 (13)	10.2 (10)	9.3 (9)	14 (12)
25 mM	At 40 °C	14.7 (13.8)	12 (11)	12.5 (11)	16.6 (15)
37.5 mM		17 (15)	13.8 (13)	14 (12.3)	18.5 (15.3)
50 mM		18.2 (17.4)	15 (14.3)	17.4 (15.6)	20.8 (20)
12.5 mM		10 (8)	8 (8)	8.6 (7.2)	12 (10)
25 mM	At 70 °C	11.5 (11)	10 (8.5)	9.7 (9)	12.8 (11.2)
37.5 mM		13 (10.3)	10.5 (10)	10.5 (9.2)	14 (12.6)
50 mM		14.8 (14)	12 (10)	11.5 (11)	17.2 (15.4)
12.5 mM		8 (7)	7.6 (6.3)	7 (6)	10.6 (10)
25 mM	At 100 °C	9 (8.4)	8.2 (7)	8.5 (7)	11 (9.3)
37.5 mM		11 (10.5)	8.8 (8)	9 (8.2)	11.8 (11)
50 mM		11.8 (10)	10 (8.8)	10 (9)	13.5 (12.6)

[1] ZI (zone of inhibitions). [2] Values in parentheses indicate durability of antimicrobial activity after 10 washing cycles.

On the other hand, the antimicrobial activity of the Se-dyed samples showed a slight decrease with an increase in the dyeing temperatures, which may be attributed to the significant impact of heat treatment on the size, structure and bioactivity of Se-NPs. Typically, smaller NPs have higher antimicrobial activity; this phenomenon can be explained by the small NPs having a larger surface area than larger NPs, which can extremely increase the production of highly reactive oxygen species (ROS), which severely damaged and inactivated the essential biomolecules, including proteins, DNA and lipids [57,58]

Moreover, subjecting the Se-dyed samples to up to 10 cycles of consecutive laundry cycles according to AATCC Test Method 61(2A)-1996 led to a non-sense decrement in their antimicrobial properties. Colored wool fabrics with low amounts of Se-NPs were enough to achieve excellent antimicrobial activity.

On the other hand, Adomaviciute et al. (2016) revealed that no antimicrobial activity was observed for the PVP solution (8%) against *Staphylococcus aureus*, *Bacillus cereus*, *Escherichia coli* and *Candida albicans* [59]. However, the PVP concentration required for Se-NPs preparation did not exceed 1.5% in this study, indicating that the antimicrobial activity mainly corresponded to Se-NPs.

These results demonstrated the activity of Se-NPs for ingrain dyed wool fabrics against the microbial pathogens and confirmed the high efficiency of the current processing in the acquisition of wool fabrics' bio-functionality, in addition to the ingrain coloration of fabrics.

3.2.2. UV-Protection Properties

Skin diseases, including allergies, premature skin aging, sunburn, and skin cancer, can occur due to skin exposure to UV irradiation. A great number of approaches were done to protect human skin against the harmful effects of UV irradiation using nanoparticles [60–62]. Therefore, the UV-blocking properties of the wool fabrics dyed with Se-NPs were evaluated. Table 2 showd the transmittance values of UV light in the range of UV-A (315–400 nm) and UV-B (280–315 nm) ranges and the UV protection factor (UPF) values of the undyed wool fabric and the wool fabrics dyed with Se-NPs. The average transmittance values of the wool fabrics were decreased obviously after wool fabrics were dyed using Se-NPs, which confirmed that the Se-NPs prominently enhanced the UV-blocking activity of wool fabrics. Both transmittance values in the UV-A and UV-B regions decreased with increasing concentrations of Se-NPs on the wool fabrics.

Table 2. UV protection properties of Se-NPs dyed wool fabrics at different temperatures.

Se-NPs conc.	Temp. °C	UV Protection Properties		
		UV-A Transmittance	UV-B Transmittance	UPF
Blank wool		10.43	1.72	33.46
12.5 mM	At room temp.	2.33	1.58	62.5 (60.2) [1]
25 mM		1.51	1.37	81.8 (80)
37.5 mM		1.46	0.83	120 (117)
50 mM		1.21	0.36	255 (250)
12.5 mM	At 40 °C	1.44	1.23	90 (87)
25 mM		1.32	0.96	130 (125.6)
37.5 mM		1.04	0.54	189.5 (180.7)
50 mM		0.76	0.21	274 (270)
12.5 mM	At 70 °C	0.71	0.88	133.5 (130.2)
25 mM		0.52	0.73	176 (170.5)
37.5 mM		0.35	0.61	258.7 (255)
50 mM		0.27	0.42	321 (320)
12.5 mM	At 100 °C	0.59	0.77	170.2 (166)
25 mM		0.5	0.65	210 (204.8)
37.5 mM		0.41	0.32	266.9 (260)
50 mM		0.08	0.06	452.6 (447.5)

[1] Values in parentheses indicate durability of UV protection properties after 10 washing cycles.

The UPF refers to the fabric efficiency to block out UV irradiation from passing through and reaching the skin. The UPF value of the undyed wool fabrics was also measured to be 33.46 (Table 2). From the results obtained, it was obvious that the dyeing process with Se-NPs increased the UPF value of wool fabrics, as shown in Figure 4 and Table 2. The results confirmed that the Se-NP dyed wool fabrics provide excellent UV protection, which increased by increasing the Se concentrations.

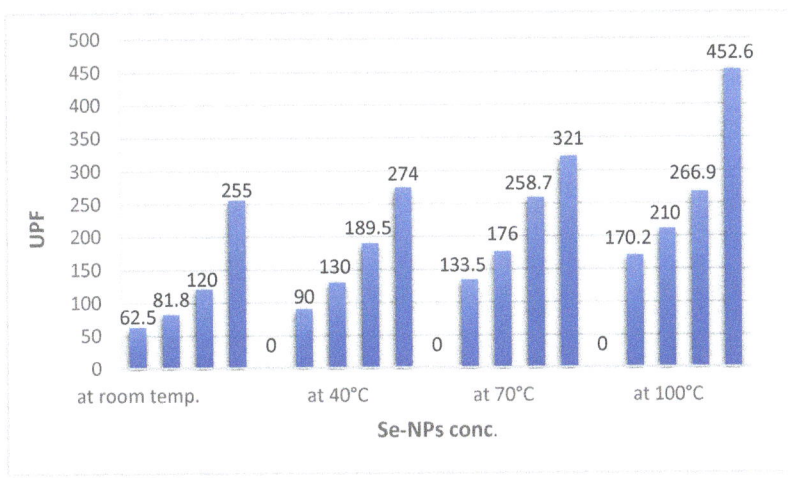

Figure 4. UV protection properties of Se-dyed wool fabrics at different temperatures.

Additionally, the influence of dyeing temperature on the UV protection properties of wool fabrics was examined. The UPF value of the wool fabrics were increased dramatically, as shown in Figure 4, as well as the UV light transmittance values were decreased with an increase in dyeing temperature. This may be due to the larger size of Se particles because of high temperature coloration; it could be considered that enlarged particles have a better chance to reflect more radiation [45].

Furthermore, the results in Table 2 also showed that repeated laundering for up to 10 cycles of the Se-dyed samples, evaluated according to AATCC Test Method 61(2A)-1996, caused a slight decrease in the imparted UV-protection functionality, expressed as UPF values; however, it was still rated as excellent values.

3.3. Color Characteristics and Color Fastness of Se-Dyed Wool Fabrics

At the end of the dyeing process, the color of the wool fabrics was changed, and the fabric color reflected the color of the impregnated Se-NPs, as shown in Figures 5 and 6. Color characteristics were quantified by estimating the color space in terms of L*, a*, b* and color strength (K/S). In this system, L* refers to lightness/darkness values from 100 to 0 representing white to black, a* values run from negative (green) to positive (red) and b* values run from negative (blue) to positive (yellow). The detected data are represented in Table 3. An illustration of the color data can be pointed in the following: dyeing at room temperature and 40 °C resulted in linear decrement in lightness values with increment in Se content, which was reflected in the color acquired by the incorporation of Se-NPs on wool fabrics. The lightness results were reduced from 79.77 to 59.14 at room temperature and from 75.56 to 42.32 for dyeing at 40 °C by increment in Se content from 12.5 to 50 mM.

Red/green values (a*) were found to be raised in a positive direction by the increment in Se content, which might have been attributed to acquiring the red color by ingrain clustering of Se-NPs, and the yellow/blue ratio (b*) was slightly increased by increasing the content of Se, reflecting the redness/yellowness color of the fabrics, as shown in Figure 5, which increased with Se content.

Figure 5. Photographs of Se-dyed wool fabrics with different concentrations of synthesized Se-NPs at room temperature and 40 °C.

Figure 6. Photographs of Se-dyed wool fabrics with different concentrations of synthesized Se-NPs at 70 °C and 100 °C.

Table 3. Color characteristics for Se-dyed wool fabrics at different temperatures.

Se-NPs Conc.	Temp. °C	L*	a*	b*	c*	H*	K/S
Blank wool		86.87	0.96	12.33	12.37	85.56	1.1
12.5 mM		79.77	7.83	20.2	21.67	68.81	2.21 (2)[1]
25 mM	Room temp.	67.81	18.84	26.05	32.15	60.12	3.71 (3.5)
37.5 mM		68.77	17.97	28.98	34.1	58.19	4.68 (4.3)
50 mM		59.14	24.52	29.26	37.41	49.05	6.63 (6.4)
12.5 mM		75.56	11.47	21.36	24.24	61.77	2.53 (2.5)
25 mM	40 °C	64.14	22.06	27.21	32.75	57.66	3.34 (3.2)
37.5 mM		66.9	23.85	29.28	32.15	51.86	4.23 (4)
50 mM		42.32	27.34	30.12	39.77	44.24	6.31 (6)
12.5 mM		42.16	12.82	15.1	19.81	49.67	9.8 (9.5)
25 mM	70 °C	42.76	18.03	21.51	28.07	50.02	12.7 (12.3)
37.5 mM		42.26	24.63	28.37	37.57	49.04	15.3 (15)
50 mM		36.03	34.3	27.04	42.25	49.99	16.3 (16)
12.5 mM		33.21	5.76	9.09	10.76	57.63	13.8 (13.3)
25 mM	100 °C	29.27	13.56	13.56	18.44	42.63	18.2 (18)
37.5 mM		28.42	16.96	16.96	23.01	42.53	21.7 (21.4)
50 mM		28.43	17.06	17.06	23.15	42.53	22.1 (22)

* Values in parentheses indicate color strength (K/S) properties after 10 washing cycles.

By increasing the dyeing temperature from 40 °C to 70 °C and 100 °C, the wool fabrics became red or brown, as indicated by the positive a* and b* values. The color of the wool fabrics darkened and changed to brown as the Se content and dyeing temperature increased, as shown in Figure 6, which indicated that temperature had a visible influence on the colors of Se-NP dyed wool fabrics, and this might be due to the increase on the sizes of Se-NPs when the solution was subjected to heat treatment [58]. The color strength (K/S) increased with the increase of dyeing temperature and Se-NP concentration. The maximum intensity of K/S was recorded in the wavelength range of 360–390 nm.

From the observations, it could be concluded that the dyeing process of wool fabric using synthesized Se-NPs resulted in the change of the fabric's color from the original creamy-white color to a yellowish red (orange) color, varying to a brownish color, according to the increment in dyeing temperature and the concentration of Se-NPs. The significant increase in color depth with increasing Se content confirmed the main responsibility of Se-NPs in the ingrain coloration of wool fabrics due to the SPR effects of Se-NPs.

Moreover, Table 3 revealed that the color depth (K/S) of the Se-dyed samples were still high after 10 washing cycles of successive laundry cycles. This indicates that the Se-NPs were still tightly loaded and fixed onto the simultaneous functional dyeing of the wool fabric surface. This might be due to the electrostatic interaction between wool fibers, which protonated to carry positive charges and Se-NPs carrying negative charges, leading to the coloration of fibers.

The washing, rubbing and light color fastness of dyed wool fabrics with Se-NPs at different dyeing temperatures were evaluated according to AATCC standard test methods (61–1972), (8–1972) and (16A–1972), respectively.

The results represented in Table 4 showed washing, rubbing and light fastness properties for all Se-dyed samples. Both Se-dyed wool fabrics at room temperature and 40°C exhibited excellent results for washing and rubbing fastness, which agreed with the K/S data after the washing process, while the light fastness showed a very good rating of 4/5.

Table 4. Fastness properties of Se-dyed wool fabrics at different temperatures.

Se-NPs Conc.	Temp. °C	WF [1] Alt.	St.	RF [2] Wet	Dry	LF [3]
12.5 mM	Room temp.	5 (5) [4]	5 (5)	5 (5)	5 (5)	4/5 (4/5)
25 mM		5 (5)	5 (5)	5 (5)	5 (5)	4/5 (4/5)
37.5 mM		5 (5)	5 (5)	5 (5)	5 (5)	4/5 (4/5)
50 mM		5 (5)	5 (5)	5 (5)	5 (5)	4/5 (4/5)
12.5 mM	40 °C	5 (5)	5 (5)	5 (5)	5 (5)	4/5 (4/5)
25 mM		5 (5)	5 (5)	5 (5)	5 (5)	4/5 (4/5)
37.5 mM		5 (5)	5 (5)	5 (5)	5 (5)	4/5 (4/5)
50 mM		5 (5)	5 (5)	5 (5)	5 (5)	4/5 (4/5)
12.5 mM	70 °C	4/5 (4/5)	4/5 (4/5)	4/5 (4)	4/5 (4/5)	4 (4)
25 mM		4/5 (4/5)	4/5 (4/5)	4 (4)	4/5 (4/5)	4 (4)
37.5 mM		4/5 (4/5)	4/5 (4/5)	4 (4)	4/5 (4/5)	4 (4)
50 mM		4/5 (4/5)	4/5 (4/5)	4 (4)	4/5 (4/5)	4 (4)
12.5 mM	100 °C	4/5 (4)	4/5 (4)	3/4 (3)	4 (4)	4 (4)
25 mM		4/5 (4)	4/5 (4)	3/4 (3)	4 (4)	4 (4)
37.5 mM		4/5 (4)	4/5 (4)	3/4 (3)	4 (4)	4 (4)
50 mM		4/5 (4)	4/5 (4)	3/4 (3)	4 (4)	4 (4)

[1] wash fastness. [2] rubbing fastness. [3] light fastness. [4] Values in parentheses indicate durability of fastness properties after 10 washing cycles.

On the other hand, wool samples dyed at high temperatures (70 and 100 °C) exhibited lower color fastness properties. While washing fastness had a very good rating in both color alteration and staining, the rubbing fastness showed good results, and the wet one had a rate of 4 at 70 °C to 3/4 at 100 °C. In addition, the light fastness exhibited a very slight color fading rate of 4. It is well known that the larger the colorant (in size), the poorer the fastness for the colored fabrics. Therefore, the moderate fastness properties of dyed wool fabrics at a high temperature could be attributed to the enlarged size of Se-NPs onto the fabric surface, caused by heat treatment and, consequently, easy leaching [45,58]

Moreover, a washing durability test was conducted according to AATCC Test Method 61(2A)-1996, and the results indicated that the fastness properties of all Se-dyed wool fabrics were still high, even after 10 cycles of consecutive laundry cycles. These results demonstrated that the wool fabrics dyed with synthesized Se-NPs have very good color fastness properties.

3.4. Mechanical Properties of Se-Dyed Wool Fabrics

Tensile strength (maximum load N/cm^2) and elongation at break (maximum strain %) were both evaluated for blank and dyed wool fabrics to give an indication for the change in mechanical properties of wool fabrics after dyeing with Se-NPs at a high temperature (100 °C). The results are reported in Table 5. The tensile strength as well as the elongation at break slightly decreased after the dyeing process.

Table 5. Mechanical properties of blank and Se-dyed wool fabrics at different temperatures.

Samples	Mechanical Properties	
	Tensile Strength (N/cm^2)	Elongation (%)
Blank wool	356.8	26.95
Se-dyed wool at RT.	330.8	25.54
40 °C	314.5	23.22
70 °C	284.4	20.22
100 °C	276	19.94
120 °C	176.6	13.88

Tensile strength decreased from 356.8 N/cm^2 for the blank wool to 176.6 N/cm^2 and elongation at break for the blank sample also decreased from 26.95% to 13.88% for the applied concentration of 50 mM Se-NPs at 120 °C. This decrease in mechanical properties is probably related to high temperature dyeing processes, which could damage the chemical structure of the wool fiber, which means that 100 °C is the maximum temperature for dyeing wool fabrics with Se-NPs.

All experimental findings clearly show the great efficiency of the proposed dyeing process in terms of color fastness combined with improved functionalities, such as antimicrobial activity and UV-blocking properties. These results lay the groundwork for a durable dyeing of different kinds of fabrics by use of the simple impregnation with colloidal solutions of metallic nanoparticles prepared with a sustainable method. This represents an important result, considering that, today, the main worldwide pollution of white waters is produced by textile treatments and dyes. Furthermore, the pollution produced by the textile industry, employing a large consumption of water and the use of harmful dyes (often azo dyes), is harmful for the environment and for human health, with a huge negative impact on the planet. Often, wastewater is not adequately treated to remove pollutants before being discharged into the environment. Very recently, specific, and targeted treatments have been developed as a suitable method for the removal of azo dyes from textile effluents. Many recent studies refer to the use of photocatalytic degradation under visible light, the employment of non-living cells of marine microalgae (i.e., *Nannochloropsis oceanica*) and other waste materials, with a careful optimization of their response surface methodology, as a suitable and quite valuable wastewater treatment [63–65].

4. Conclusions

The present study has reported a one step process for the multi-functionalization of dyeing wool fabrics using a colloidal solution of Se-NPs. All Se-NP concentrations were adsorbed onto wool fabrics even though the dyeing temperatures were different, which was confirmed by the color fastness properties and color strength (K/S). The obtained colored wool fabrics had a stable bright color ranging from light orange to a dark orange color according to the concentration of the prepared Se-NPs as well as the dyeing temperatures. Moreover, the wool fabrics with Se-NPs exhibited significant antimicrobial activity and excellent UV protection properties with excellent washing durability up to 10 laundering cycles. XRD, SEM and EDX analyses revealed that the Se-NP solution was effectively absorbed on wool fabric surfaces. Furthermore, a high dyeing temperature led to a change in the color characteristics from a dark orange color to dark brownish color, especially at 100 °C. However, the UV-protection properties were not affected by high temperature dyeing; the antimicrobial properties slightly decreased, and the mechanical properties of the dyed wool fabrics also decreased when the dyeing temperature was increased. The study concluded that the Se-wool dyeing process at a low temperature (at room temperature or 40 °C) was better than that of high temperature dyeing. The proposed dyeing process of wool fabrics with Se-NPs demonstrated that the dyeing of fibers based on nanoparticles may facilitate the functionalization of fibrous materials better than the traditional dyestuffs that require energy and an increased temperature for dyeing. This work will open the way to the use of opportune nanoparticles for the functionalization of different kinds of surfaces, to obtain implemented mechanical properties (i.e., color, sensing and antibacterial).

Author Contributions: T.A.E. conceived the original idea and participated in the manuscript drafting. S.R. was responsible for the dyeing process using the prepared Se-NPs and evaluation of the characteristics of the dyed wool fabrics. She discussed the obtained results with T.A.E. and M.R.P. as well as participated in the manuscript writing. K.S.-A. prepared Se-NPs and interpreted the obtained results from SEM and XRD analyses and antimicrobial activity test as well as participated in the manuscript writing. M.R.P. put a lot of effort into reviewing the manuscript and took charge of all correspondence. All authors have read and agreed to the published version of the manuscript.

Funding: The authors are grateful to the P.O. FESR SICILIA 2014/2020 n. 08CL4120000131-SETI (Sicilia Eco Tecnologie Innovative- n. 08CL4120000131) project for funding.

Institutional Review Board Statement: Not applicable.

Informed Consent Statement: Not applicable.

Data Availability Statement: The data presented in this study are available on request from the corresponding author.

Acknowledgments: MURST: CNR and MIUR are gratefully acknowledged for financial support.

Conflicts of Interest: The authors declare no conflict of interest.

References

1. Yetisen, A.K.; Qu, H.; Manbachi, A.; Butt, H.; Dokmeci, M.R.; Hinestroza, J.P.; Skorobogatiy, M.; Khademhosseini, A.; Yun, S.H. Nanotechnology in textiles. *ACS Nano* **2016**, *10*, 3042–3068. [CrossRef] [PubMed]
2. Abou Elmaaty, T.; Elsisi, H.G.; Elsayad, G.M.; Elhadad, H.H.; Sayed-Ahmed, K.; Plutino, M.R. Fabrication of New Multifunctional Cotton/Lycra Composites Protective Textiles through Deposition of Nano Silica Coating. *Polymers* **2021**, *13*, 2888. [CrossRef]
3. Castellano, A.; Colleoni, C.; Iacono, G.; Mezzi, A.; Plutino, M.R.; Malucelli, G.; Rosace, G. Synthesis and characterization of a phosphorous/nitrogen based sol-gel coating as a novel halogen- and formaldehyde-free flame retardant finishing for cotton fabric. *Polym. Degrad. Stab.* **2019**, *162*, 148–159. [CrossRef]
4. Plutino, M.R.; Colleoni, C.; Donelli, I.; Freddi, G.; Guido, E.; Maschi, O.; Mezzi, A.; Rosace, G. Sol-gel 3-glycidoxypropyltriethoxysilane finishing on different fabrics: The role of precursor concentration and catalyst on the textile performances and cytotoxic activity. *J. Colloid Interface Sci.* **2017**, *506*, 504–517. [CrossRef] [PubMed]
5. Rosace, G.; Guido, E.; Colleoni, C.; Brucale, M.; Piperopoulos, E.; Milone, C.; Plutino, M.R. Halochromic resorufin-GPTMS hybrid sol-gel: Chemical-physical properties and use as pH sensor fabric coating. *Sens. Actuators B Chem.* **2017**, *241*, 85–95. [CrossRef]
6. Dong, J.; Luo, S.; Ning, S.; Yang, G.; Pan, D.; Ji, Y.; Liu, C. MXene-Coated Wrinkled Fabrics for Stretchable and Multifunctional Electromagnetic Interference Shielding and Electro/Photo-Thermal Conversion Applications. *ACS Appl. Mater. Interf.* **2021**, *13*, 60478–60488. [CrossRef] [PubMed]
7. Plutino, M.R.; Guido, E.; Colleoni, C.; Rosace, G. Effect of GPTMS functionalization on the improvement of the pH-sensitive methyl red photostability. *Sens. Actuators B* **2017**, *238*, 281–291. [CrossRef]
8. Trovato, V.; Rosace, G.; Colleoni, C.; Sfameni, S.; Migani, V.; Plutino, M.R. Sol-gel based coatings for the protection of cultural heritage textiles. *IOP Conf. Ser. Mater. Sci. Eng.* **2020**, *777*, 012007. [CrossRef]
9. Ielo, I.; Galletta, M.; Rando, G.; Sfameni, S.; Cardiano, P.; Sabatino, G.; Drommi, D.; Rosace, G.; Plutino, M.R. Design, synthesis, and characterization of hybrid coatings suitable for geopolymeric-based supports for the restoration of cultural heritage. *IOP Conf. Ser. Mater. Sci. Eng.* **2020**, *777*, 012003. [CrossRef]
10. Rosace, G.; Cardiano, P.; Urzì, C.; De Leo, F.; Galletta, M.; Ielo, I.; Plutino, M.R. Potential roles of fluorine-containing sol-gel coatings against adhesion to control microbial biofilm. *IOP Conf. Ser. Mater. Sci. Eng.* **2018**, *459*, 012021. [CrossRef]
11. Grancarić, A.M.; Tarbuk, A.; Sutlović, A.; Castellano, A.; Colleoni, C.; Rosace, G.; Plutino, M.R. Enhancement of acid dyestuff salt-free fixation by a cationizing sol-gel based coating for cotton fabric. *Colloids Surf. A Physicochem. Eng. Asp.* **2021**, *612*, 125984. [CrossRef]
12. Puoci, F.; Saturnino, C.; Trovato, V.; Iacopetta, D.; Piperopoulos, E.; Triolo, C.; Bonomo, M.G.; Drommi, D.; Parisi, O.I.; Milone, C.; et al. Sol–Gel Treatment of Textiles for the Entrapping of an Antioxidant/Anti-Inflammatory Molecule: Functional Coating Morphological Characterization and Drug Release Evaluation. *Appl. Sci.* **2020**, *10*, 2287. [CrossRef]
13. Ielo, I.; Giacobello, F.; Sfameni, S.; Rando, G.; Galletta, M.; Trovato, V.; Rosace, G.; Plutino, M.R. Nanostructured Surface Finishing and Coatings: Functional Properties and Applications. *Materials* **2021**, *14*, 2733. [CrossRef] [PubMed]
14. AbouElmaaty, T.; Abdeldayem, S.A.; Ramadan, S.M.; Sayed-Ahmed, K.; Plutino, M.R. Coloration and Multi-Functionalization of Polypropylene Fabrics with Selenium Nanoparticles. *Polymers* **2021**, *13*, 2483. [CrossRef] [PubMed]
15. Ibrahim, N.A.; Eid, B.M.; El-Aziz, E.A.; Elmaaty, T.M.A.; Ramadan, S.M. Loading of chitosan–Nano metal oxide hybrids onto cotton/polyester fabrics to impart permanent and effective multifunctions. *Int. J. Biol. Macromol.* **2017**, *105*, 769–776. [CrossRef] [PubMed]
16. Ibrahim, N.A.; Eid, B.M.; El-Aziz, E.A.; Elmaaty, T.M.A.; Ramadan, S.M. Multifunctional cellulose-containing fabrics using modified finishing formulations. *RSC Adv.* **2017**, *7*, 33219–33230. [CrossRef]
17. Ibrahim, N.A.; Eid, B.M.; El-Aziz, E.A.; Elmaaty, T.M.A. Functionalization of linen/cotton pigment prints using inorganic nano structure materials. *Carbohydr. Polym.* **2013**, *97*, 537–545. [CrossRef] [PubMed]
18. Ibrahim, N.A.; Eid, B.M.; Elmaaty, T.M.A.; El-Aziz, E.A. A smart approach to add antibacterial functionality to cellulosic pigment prints. *Carbohydr. Polym.* **2013**, *94*, 612–618. [CrossRef] [PubMed]
19. El-bisi, M.; Othman, R.; Yassin, F.A. Improving antibacterial and ultraviolet properties of cotton fabrics via dual effect of nano-metal oxide and Moringa oleifera extract. *Egypt. J. Chem.* **2020**, *63*, 3441–3451. [CrossRef]
20. Ibrahim, N.A.; El-Aziz, E.A.; Eid, B.M.; Elmaaty, T.M.A. Single-stage process for bifunctionalization and eco-friendly pigment coloration of cellulosic fabrics. *J. Text. Inst.* **2016**, *107*, 1022–1029.
21. Ibrahim, N.A.; Eid, B.M.; Abdel-Aziz, M.S. Effect of plasma superficial treatments on antibacterial functionalization and coloration of cellulosic fabrics. *Appl. Surf. Sci.* **2017**, *392*, 1126–1133. [CrossRef]

22. Ibrahim, N.A.; Nada, A.A.; Eid, B.M.; Al-Moghazy, M.; Hassabo, A.G.; Abou-Zeid, N.Y. Nano-structured metal oxides: Synthesis, characterization and application for multifunctional cotton fabric. *Adv. Nat. Sci. Nanosci. Nanotechnol.* **2018**, *9*, 35014. [CrossRef]
23. Eid, B.M.; Ibrahim, N.A. Recent developments in sustainable finishing of cellulosic textiles employing biotechnology. *J. Clean. Prod.* **2021**, *284*, 124701–124722. [CrossRef]
24. Barattucci, A.; Plutino, M.R.; Faggi, C.; Bonaccorsi, P.; Monsù Scolaro, L.; Aversa, M.C. Mono- and trinuclear tripodal platinum(II) chelated complexes containing a pyridine/sulfoxide based anchoring framework. *Eur. J. Inorg. Chem.* **2013**, *19*, 3412–3420. [CrossRef]
25. Ibrahim, H.M.; Zaghloul, S.; Hashem, M.; El-Shafei, A. A green approach to improve the antibacterial properties of cellulose based fabrics using Moringa oleifera extract in presence of silver nanoparticles. *Cellulose* **2021**, *28*, 549–564. [CrossRef]
26. De Luca, G.; Bonaccorsi, P.; Trovato, V.; Mancuso, A.; Papalia, T.; Pistone, A.; Casaletto, M.P.; Mezzi, A.; Brunetti, B.; Minuti, L.; et al. Tripodal tris-disulfides as capping agents for a controlled mixed functionalization of gold nanoparticles. *New J. Chem.* **2018**, *42*, 16436–16440. [CrossRef]
27. Mehravani, B.; Ribeiro, A.I.; Zille, A. Gold Nanoparticles Synthesis and Antimicrobial Effect on Fibrous Materials. *Nanomaterials* **2021**, *11*, 1067. [CrossRef]
28. Ielo, I.; Iacopetta, D.; Saturnino, C.; Longo, P.; Galletta, M.; Drommi, D.; Rosace, G.; Sinicropi, M.S.; Plutino, M.R. Gold Derivatives Development as Prospective Anticancer Drugs for Breast Cancer Treatment. *Appl. Sci.* **2021**, *11*, 2089. [CrossRef]
29. Ielo, I.; Rando, G.; Giacobello, F.; Sfameni, S.; Castellano, A.; Galletta, M.; Drommi, D.; Rosace, G.; Plutino, M.R. Synthesis, Chemical–Physical Characterization, and Biomedical Applications of Functional Gold Nanoparticles: A Review. *Molecules* **2021**, *26*, 5823. [CrossRef]
30. Riaz, S.; Ashraf, M.; Hussain, T.; Hussain, M.T.; Rehman, A.; Javid, A.; Iqbal, K.; Basit, A.; Aziz, H. Functional finishing, and coloration of textiles with nanomaterials. *Color. Technol.* **2018**, *134*, 327–346. [CrossRef]
31. Rivero, P.J.; Goicoechea, J.; Arregui, F.J. Localized Surface Plasmon Resonance for Optical Fiber-Sensing Applications. In *Nanoplasmonics—Fundamentals and Applications*; Intech Open: London, UK, 2017; Chapter 17.
32. Wang, X.; Xu, D.; Jaquet, B.; Yang, Y.; Wang, J.; Huang, H.; Chen, Y.; Gerhard, C.; Zhang, K. Structural Colors by Synergistic Birefringence and Surface Plasmon Resonance. *ACS Nano* **2020**, *14*, 16832–16839. [CrossRef] [PubMed]
33. Tan, K.B.; Vakili, M.; Horri, B.A.; Poh, P.E.; Abdullah, A.Z.; Salamatinia, B. Adsorption of dyes by nanomaterials: Recent developments and adsorption mechanisms. *Sep. Purif. Technol.* **2015**, *150*, 229–242. [CrossRef]
34. Kafafy, H.; Helmy, H.; Zaher, A. Treatment of cotton and wool fabrics with different nanoparticles for multifunctional properties. *Egyptian J. Chem.* **2021**, *64*, 5255–5267. [CrossRef]
35. Wu, M.; Ma, B.; Pan, T.; Chen, S.; Sun, J. Silver-nanoparticle-colored cotton fabrics with tunable colors and durable antibacterial and self-healing superhydrophobic properties. *Adv. Funct. Mater.* **2016**, *26*, 569–576. [CrossRef]
36. Razmkhah, M.; Montazer, M.; Rezaie, A.B.; Rad, M.M. Facile technique for wool coloration via locally forming of nano selenium photocatalyst imparting antibacterial and UV protection properties. *J. Ind. Eng. Chem.* **2021**, *101*, 153–164. [CrossRef]
37. Abdelrahman, M.S.; Nassar, S.H.; Mashaly, H.; Mahmoud, S.; Maamoun, D.; El-Sakhawy, M.; Khattab, T.; Kamel, S. Studies of polylactic acid and metal oxide nanoparticles-based composites for multifunctional textile prints. *Coatings* **2020**, *10*, 58. [CrossRef]
38. Johnston, J.H.; Lucas, K.A. Nanogold synthesis in wool fibres: Novel colourants. *Gold Bull.* **2011**, *44*, 85–89. [CrossRef]
39. Rehan, M.; Elshemy, N.S.; Haggag, K.; Montaser, A.S.; Ibrahim, G.E. Phytochemicals and volatile compounds of peanut red skin extract: Simultaneous coloration and in situ synthesis of silver nanoparticles for multifunctional viscose fibers. *Cellulose* **2020**, *27*, 9893–9912. [CrossRef]
40. Tang, B.; Yao, Y.; Chen, W.; Chen, X.; Zou, F.; Wang, X. Kinetics of dyeing natural protein fibers with silver nanoparticles. *Dye. Pigment.* **2018**, *148*, 224–235. [CrossRef]
41. Jafari-Kiyan, A.; Karimi, L.; Davodiroknabadi, A. Producing colored cotton fabrics with functional properties by combining silver nanoparticles with nano titanium dioxide. *Cellulose* **2017**, *24*, 3083–3094. [CrossRef]
42. Tang, B.; Li, J.; Hou, X.; Afrin, T.; Sun, L.; Wang, X. Colorful and antibacterial silk fiber from anisotropic silver nanoparticles. *Ind. Eng. Chem. Res* **2013**, *52*, 4556–4563. [CrossRef]
43. Tang, B.; Sun, L.; Kaur, J.; Yu, Y.; Wang, X. In-situ synthesis of gold nanoparticles for multifunctionalization of silk fabrics. *Dye. Pigment.* **2014**, *103*, 183–190. [CrossRef]
44. Yao, Y.; Tang, B.; Chen, W.; Sun, L.; Wang, X. Sunlight-induced coloration of silk. *Nanoscale Res. Lett.* **2016**, *11*, 293. [CrossRef] [PubMed]
45. Emam, H.E.; Rehan, M.; Mashaly, H.M.; Ahmed, H.B. Large scaled strategy for natural/synthetic fabrics functionalization via immediate assembly of AgNPs. *Dye. Pigment.* **2016**, *133*, 173–183. [CrossRef]
46. Tang, B.; Sun, L.; Li, J.; Kaur, J.; Zhu, H.; Qin, S.; Yao, Y.; Chen, W.; Wang, X. Functionalization of bamboo pulp fabrics with noble metal nanoparticles. *Dye. Pigment.* **2015**, *113*, 289–298. [CrossRef]
47. Tang, B.; Yao, Y.; Li, J.; Qin, S.; Zhu, H.; Kaur, J.; Chen, W.; Sun, L.; Wang, X. Functional application of noble metal nanoparticles in situ synthesized on ramie fibers. *Nanoscale Res. Lett.* **2015**, *10*, 366. [CrossRef] [PubMed]
48. Mowafi, S.; Rehan, M.; Mashaly, H.M.; Abou El-Kheir, A.; Emam, H.E. Influence of silver nanoparticles on the fabrics functions prepared by in-situ technique. *J. Text. Inst.* **2017**, *108*, 1828–1839. [CrossRef]
49. Hong, Z. Preparation and antioxidant properties of selenium nanoparticles-loaded chitosan microspheres. *Int. J. Nanomed.* **2017**, *12*, 4527–4539.

50. Zhai, X.; Zhang, C.; Zhao, G.; Stoll, S.; Ren, F.; Leng, X. Antioxidant capacities of the selenium nanoparticles stabilized by chitosan. *J. Nanobiotechnology* **2017**, *15*, 4. [CrossRef]
51. Nastulyavichus, A.; Kudryashov, S.; Smirnov, N.; Saraeva, I.; Rudenko, A.; Tolordava, E.; Ionin, A.; Romanova, Y.; Zayarny, D. Antibacterial coatings of Se and Si nanoparticles. *Appl. Surf. Sci* **2019**, *469*, 220–225. [CrossRef]
52. Huang, X.; Chen, X.; Chen, Q.; Yu, Q.; Sun, D.; Liu, J. Investigation of functional selenium nanoparticles as potent antimicrobial agents against superbugs. *Acta Biomater.* **2016**, *30*, 397–407. [CrossRef] [PubMed]
53. Yip, J.; Liu, L.; Wong, K.H.; Leung, P.H.M.; Yuen, C.W.M.; Cheung, M.C. Investigation of antifungal and antibacterial effects of fabric padded with highly stable selenium nanoparticles. *J. Appl. Polym. Sci.* **2014**, *131*, 8886–8893. [CrossRef]
54. Elmaaty, T.A.; El-Nagare, K.; Raouf, S.; Abdelfattah, K.; El-Kadi, S.; Abdelaziz, E. One-step green approach for functional printing and finishing of textiles using silver and gold NPs. *RSC Adv.* **2018**, *8*, 25546–25557. [CrossRef]
55. Elmaaty, T.A.; Raouf, S.; Sayed-Ahmed, K. Novel one step printing and functional finishing of wool fabric using selenium nanoparticles. *Fiber Polym.* **2019**, *21*, 1983–1991. [CrossRef]
56. Vieira, A.P.; Stein, E.M.; Andreguetti, D.X.; Cebrián-Torrejón, G.; Doménech-Carbó, A.; Colepicolo, P.; Ferreira, A.M.D.C. "Sweet Chemistry": A Green Way for Obtaining Selenium Nanoparticles Active against Cancer Cells. *J. Braz. Chem. Soc.* **2017**, *28*, 2021–2027. [CrossRef]
57. Slavin, Y.N.; Asnis, J.; Häfeli, U.O.; Bach, H. Metal nanoparticles: Understanding the mechanisms behind antibacterial activity. *J. Nanobiotechnology* **2017**, *15*, 65. [CrossRef] [PubMed]
58. Zhang, J.; Taylor, E.W.; Wan, X.; Peng, D. Impact of heat treatment on size, structure, and bioactivity of elemental selenium nanoparticles. *Int. J. Nanomed.* **2012**, *7*, 815. [CrossRef]
59. Adomaviciute, E.; Stanys, S.; Žilius, M.; Juškaite, V.; Pavilonis, A.; Briedis, V. Formation and biopharmaceutical characterization of electrospun PVP mats with propolis and silver nanoparticles for fast releasing wound dressing. *Biomed. Res. Int.* **2016**, *2016*, 4648287. [CrossRef]
60. Lu, Z.; Mao, C.; Meng, M.; Liu, S.; Tian, Y.; Yu, L.; Sun, B.; Li, C.M. Fabrication of CeO_2 nanoparticle-modified silk for UV protection and antibacterial applications. *J. Colloid Interface Sci.* **2014**, *435*, 8–14. [CrossRef] [PubMed]
61. Khan, M.Z.; Baheti, V.; Ashraf, M.; Hussain, T.; Ali, A.; Javid, A.; Rehman, A. Development of UV protective, superhydrophobic and antibacterial textiles using ZnO and TiO_2 nanoparticles. *Fibers Polym.* **2018**, *19*, 1647–1654. [CrossRef]
62. Chitichotpanya, P.; Pisitsak, P.; Chitichotpanya, C. Sericin—Copper-functionalized silk fabrics for enhanced ultraviolet protection and antibacterial properties using response surface methodology. *Text. Res. J.* **2019**, *89*, 1166–1179. [CrossRef]
63. Zuorro, A.; Lavecchia, R.; Monaco, M.M.; Iervolino, G.; Vaiano, V. Photocatalytic Degradation of Azo Dye Reactive Violet 5 on Fe-Doped Titania Catalysts under Visible Light Irradiation. *Catalysts* **2019**, *9*, 645. [CrossRef]
64. Zuorro, A.; Maffei, G.; Lavecchia, R. Kinetic modeling of azo dye adsorption on non-living cells of Nannochloropsis oceanica. *J. Environ. Chem. Eng.* **2017**, *5*, 4121–4127. [CrossRef]
65. Panusa, A.; Zuorro, A.; Lavecchia, R.; Marrosu, G.; Petrucci, R. Recovery of Natural Antioxidants from Spent Coffee Grounds. *J. Agric. Food Chem.* **2013**, *61*, 4162–4168. [CrossRef]

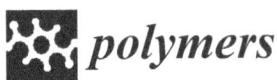

Article

Soybean Oil-Based Biopolymers Induced by Nonthermal Plasma to Enhance the Dyeing of Para-Aramids with a Cationic Dye

Caleb Metzcar [1], Xiaofei Philip Ye [2,*], Toni Wang [1] and Christopher J. Doona [3,4]

[1] Department of Food Science, The University of Tennessee, Knoxville, TN 37996, USA; cmetzcar@vols.utk.edu (C.M.); twang46@utk.edu (T.W.)
[2] Department of Biosystems Engineering and Soil Science, The University of Tennessee, Knoxville, TN 37996, USA
[3] U.S. Army Combat Capabilities Development Command—Soldier Center, Natick, MA 01760, USA; doonac@mit.edu
[4] Massachusetts Institute of Technology—Institute for Soldier Nanotechnologies, 77 Massachusetts Ave NE47-4F, Cambridge, MA 02139, USA
* Correspondence: xye2@utk.edu; Tel.: +1-865-974-7129; Fax: +1-865-974-4514

Citation: Metzcar, C.; Ye, X.P.; Wang, T.; Doona, C.J. Soybean Oil-Based Biopolymers Induced by Nonthermal Plasma to Enhance the Dyeing of Para-Aramids with a Cationic Dye. *Polymers* **2022**, *14*, 628. https://doi.org/10.3390/polym14030628

Academic Editor: Tarek M. Abou Elmaaty

Received: 18 January 2022
Accepted: 4 February 2022
Published: 6 February 2022

Publisher's Note: MDPI stays neutral with regard to jurisdictional claims in published maps and institutional affiliations.

Copyright: © 2022 by the authors. Licensee MDPI, Basel, Switzerland. This article is an open access article distributed under the terms and conditions of the Creative Commons Attribution (CC BY) license (https://creativecommons.org/licenses/by/4.0/).

Abstract: To overcome the recalcitrance of para-aramid textiles against dyeing, this study demonstrated that increasing the functionalities of soybean oil applied to the surface of para-aramids followed by a nonthermal plasma (NTP) treatment improved the dyeing color strength compared with the use of soybean oil alone, and that dyeing occurred through covalent bonding. Particularly, compared with the pretreatment using soybean oil that obtained the highest color strength of 3.89 (as K/S value determined from spectral analysis of the sample reflectance in the visible range), the present pretreatments with either acrylated epoxidized soybean oil (AESO) or a mixture of acrylic acid and soybean oil (AA/Soy) achieved K/S values higher than nine (>9.00). The NTP treatment, after the AESO or AA/Soy pretreatment, was essential in inducing the formation of a polymerized network on the surface of para-aramids that bonded the dye molecules and generating covalent bonds that anchored the polymerized network to the para-aramids, which is difficult to achieve given the high crystallinity and chemical inertness of para-aramids. As an important economic consideration, the sequential experimentation method demonstrated that a simple mixture of AA/Soy could replace the expensive AESO reagent and render a comparable performance in dyeing para-aramids. Among the auxiliary additives tested with the AESO and AA/Soy pretreatments followed by NPT treatment in this study, Polysorbate 80 as a surfactant negatively affected the dyeing, benzyl alcohol as a swelling agent had minimal effect, and NaCl as an electrolyte showed a positive effect. The dyeing method developed in this study did not compromise the strength of para-aramids.

Keywords: para-aramid; nonthermal plasma; ambient air; soybean oil; acrylic acid; acrylated epoxidized soybean oil; cationic dye

1. Introduction

The invention of aramids with commercial names such as Kevlar® (para-aramid) and Nomex® (meta-aramid) brought about a synthetic fiber with high strength-to-weight ratio, low elongation to break, superior heat and flame resistance, high cut resistance, and excellent ballistic properties [1]. Since the first commercial use of Kevlar® in the early 1970s as a replacement for steel in racing tires, aramid materials have been used in an increasing number of diverse applications. Examples include textiles for protective clothing (e.g., flame-resistant apparel) and body armor (e.g., bullet-proof vests, helmets, and puncture-resistant correctional wear), sportswear, and reinforced composites (e.g., brake pads, gaskets, hot-air filters, industrial belts and ropes, and strength member in fiber optics) [2]. It is highly desirable to incorporate aramids into additional applications for protective clothing, but these efforts are complicated by the difficulty associated with durably dyeing or printing aramids.

Between the two major types of aramids, meta-aramids and para-aramids, the meta-aramid fibers consist of poly(m-phenylene isophthalamide) that binds via meta-linked aromatic rings to result in a semi-crystalline fiber with the molecular chain oriented along the fiber axis, while the building-blocks of para-aramids are poly(p-phenylene terephthalamide) with stiff para-linked aromatic rings and densely arranged hydrogen bond donors and acceptors throughout their backbones [3]. This inherent molecular rigidity of para-aramids, combined with strong intermolecular hydrogen bonding interactions, enables the molecules to achieve excellent alignment with their neighbors, resulting in a highly anisotropic unit cell consisting of covalent bonds, hydrogen bonds, and van der Waals interactions along each fundamental axis, forming a highly crystalline structure [4]. The high degree of crystallinity and chemical inertness makes it difficult to dye para-aramids with conventional dyeing methods, because para-aramids cannot entrap or bind dye molecules.

There are a number of methods reported for dyeing meta-aramids, but few methods reported for dyeing para-aramids (especially continuous filament para-aramids) without using harsh chemicals that can damage the dyed materials and sacrifice mechanical strength [5–7]. Typically, aramid filaments/yarns are dyed to a single color using solution dyeing methods, a process in which acid-tolerant colorant is added to the polymer dope at the time of aramid filament production, and the dyed filaments are subsequently woven into usable fabrics with only a limited number of color choices available by these methods [8]. While it is highly desirable to incorporate aramids into new applications for protective clothing or outerwear, for example, the difficulty in durably dyeing or printing aramids is a barrier to using aramids in these applications. New methods of dyeing para-aramids that avoid the use of harsh or environmentally unsafe chemicals would be advantageous for these applications.

Our previous studies [9,10] demonstrated that pretreatment with soybean oil followed by nonthermal plasma (NTP) treatment enabled dyeing para-aramids to significantly high color strength without other chemical additives; and this new method is compatible with both a disperse dye and a basic dye, showing the potential of this method replacing the environmentally unfriendly chemicals in current dyeing practices with renewable, environmentally friendly materials to improve the dyeing of para-aramid fabrics. The proposed mechanism for improved dyeing is that the soybean oil diffuses onto woven fabric and adsorbs onto the surfaces of yarns and fibers, and the subsequent NTP treatment induces the formation of a polymerized network in situ, enabling dyeing to a higher color strength. The color strength in term of K/S value, determined from spectral analysis of the sample reflectance in the visible range, increased to 3.89 from ~1 of the untreated samples. Along this line of sustainable dyeing method, in this study, we aim at further improving the dyeing color strength and colorfastness by deriving more functionalities of soybean oil on the surface of para-aramids.

Acrylated epoxidized soybean oil (AESO) synthesized from soybean oil has already occupied a significant share of market as a "green" alternative to petroleum-based epoxy resins, plasticizers, and pre-polymers [11]. It contains three highly reactive functionalities of double (C=C) bonds, –OH groups, and epoxy rings. The C=C bond in AESO is capable of self-polymerizing and copolymerizing with other components via a free radical initiation (including UV and NTP, which generate both UV and free radical species), forming a network with ample functionalities to bind with dyes [12]. However, the current method of soybean oil epoxidation followed by acrylate addition for AESO production is a tedious process, rendering a high price for AESO. We hypothesized that pretreating para-aramids with a simple mixture of soybean oil and acrylic acid followed by NTP treatment would induce a similar polymerized network in situ to enhance the dyeing. Importantly, acrylic acid can also be produced from renewable glycerol, which is a co-product of massive biodiesel production, mainly from soybean oil in the U.S. Acrylic acid can be produced via the intermediate of glycerol dehydration to acrolein, and this strategy has received much attention because it appears to be one of the most promising ways to valorize glycerol [13–15].

Furthermore, because the dyeing industry often uses auxiliary chemical additives of surfactants, electrolytes, and swelling agents to improve dyeing performance [16–18], the effects of Polysorbate 80 as a surfactant, sodium chloride (NaCl) as an electrolyte, and benzyl alcohol as a swelling agent, were also examined in our new method.

2. Materials and Methods

The following section describes all materials and methods used in this study. Overall, a series of experiments, designed as sequential experimentation, directed this study and demonstrated the potential for this new method to improve the dyeing of para-aramids. The experimental design and analysis were conducted using Design-Expert software (version 6, Stat-Ease, Inc., Minneapolis, MN, USA). Because many experimental factors were involved, our design of experiments emphasized the sequential use of two-level factorial designs to identify critical factors for improving dyeing, and the sequential assembly of second-order designs to elucidate the nature of the response surface in the improved formulations and process conditions.

2.1. Materials

The para-aramid fabrics used in this study are made of tightly woven 300 Denier continuous filament and were provided by the U.S. Army Development Command—Soldier Center (Natick, MA, USA). This highly crystalline material has been proven undyeable by conventional methods using either a cationic dye or a disperse dye, even with nonthermal plasma surface treatment, but were dyed using soybean oil/NTP pretreatments [9,10]. Blue cationic dye (Victoria Blue R, CAS Number 2185-86-6), Disperse Red 1 acrylate dye (CAS Number 13695-46-0), AESO (CAS Number 91722-14-4), acrylic acid (CAS Number 79-10-7), and ethyl acetate (CAS Number 141-78-6) were purchased from Sigma-Aldrich (St. Louis, MO, USA). Polysorbate 80 (a nonionic surfactant with the common name TWEEN 80 and the IUPAC name polyoxyethylene (20) sorbitan monooleate) and benzyl alcohol (a common swelling agent) were purchased from Chem Center @ Amazon.com. Commercial food-grade refined soybean oil (typically consisting of about 23% monounsaturated fat, 58% polyunsaturated fat, and 15% saturated fat) and laundry detergent (ECOS® plus stain-fighting enzymes) were purchased from a local supermarket. All materials in this study were used as-is and without further purification.

2.2. Pretreatment and NTP Treatment

Each para-aramid fabric sample was cut into approximately one square inch swatch. Depending on the experimental design, the pretreatment was conducted by submerging the samples in a solution of AESO or a mixture of acrylic acid and soybean oil (denoted as AA/Soy hereafter) in ethyl acetate in a sealed beaker for a designated time. Then, the samples were taken out, placed on a paper towel, and pressed with a roller two times to remove excessive pretreatment liquid, followed by NTP treatment if required by the experimental design.

The NTP treatment of the fabric samples was carried out using an in-house made surface dielectric barrier discharge (SDBD) apparatus consisting of two electrodes separated by a 108 × 95 mm alumina dielectric plate with a thickness of 1 mm (Figure 1). The alumina plate has an induction electrode made of a rectangular copper tape embedded in insulation tape on its top and a discharge electrode made of 17 interconnected tungsten strips on its lower surface. Teflon-coated aramids have been made wettable after a 30 s exposure to this SDBD. Compared with volume dielectric barrier discharge (VDBD), SDBD generates a higher density of micro-discharges that are limited to the surface of the sample, and thereby avoids pin-holing that VDBD caused in para-aramids by the hot electron bombardment, weakening the fibers. The feedgas for the SDBD is ambient air, and the power for the SDBD was a 9.2 kV sinusoidal high voltage source tuned to a resonance frequency of 23.2 kHz.

Para-aramid samples were placed on top of a Plexiglas platform mounted on a rotating stage, and the SDBD plate was lowered via an adjustable stand to 1 mm above and parallel to the sample to ensure uniform treatment of the sample. Surface emissions, reactive oxygen and nitrogen species (RONS), and other radicals generated in the plasma interacted with the para-aramid samples. At completion of the NTP treatment of a specified time, the power to the SBSD was turned off and the sample was removed immediately thereafter.

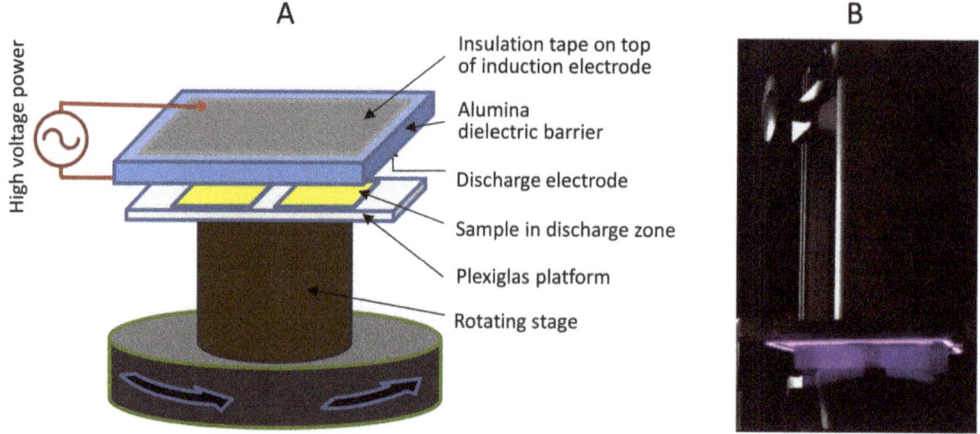

Figure 1. (**A**) Schematic of NTP treatment and (**B**) Photo image showing glowing surface discharge on top of samples in dark background.

2.3. Dyeing Experimental Procedure and Washing

The dye bath was prepared by dissolving 0.1 wt.% of Victoria Blue R dye in distilled water. Depending on the experimental design, auxiliary additives of a surfactant, electrolyte, and/or swelling agent were added to the dye bath, expressed as a mass percentage based on the solvent. Then, the dye bath liquid was transferred into vials of a combined heating/stirring system (Reacti-Therm, ThermoFisher Scientific, Waltham, MA, USA) to conduct the dyeing experiments. The Reacti-Therm system can hold up to eight vials with different dye-bath formulations and precisely control the dyeing temperature. The Reacti-Therm system was initially set at 60 °C with stirring while the samples were loaded into their respective vials, then the temperature was raised to T = 90 °C in about 30 min and held there for 1 h. Subsequently, the Reacti-Therm system was turned off, and samples were left in the vials to cool for 20 min.

After dyeing, all the samples were removed from the vials and rinsed under flowing warm tap water for 2 min followed by a cold-water rinse for another 2 min. Then, the samples were dried in a programmable convective oven starting at T = 30 °C, raising the temperature to T = 150 °C at a rate of 30 °C/min, and holding at T = 150 °C for 2 min, to fix the dye to the fabric samples. To prepare the samples for color strength analysis, the samples were washed with detergent (ECOS® plus stain-fighting enzymes) to remove the oily pretreatment materials using a home-made tumbler to simulate laundering, in accordance with the protocol described in ISO standard 105-C10:2006 [19]. After the detergent washing, the samples were rinsed and dried again in the same way as described above.

2.4. Color Strength Analysis

The color strength of each dyed sample was quantified by measuring its spectral reflectance (R in %) in the visible range using a spectrophotometer (SPECTRO 1, Variable

Inc., Chattanooga, TN, USA). The color strength was calculated at the wavelength of maximum absorbance for Victoria Blue R (λ = 615 nm) using the K/S value defined by the Kubelka–Munk equation (Equation (1)) that relates R with sample absorption (K) and scattering characteristics (S) [20].

$$K/S = \frac{(1 - 0.01R)^2}{2(0.01R)} \quad (1)$$

The Kubelka–Munk equation is used in formulating colors for the textile, paper, and coatings industries. For these applications, it is assumed that the absorption (K) of light depends on the properties of the colorant, and the scattering (S) of a dye or pigment depends on the properties of the substrate or opacifier. The K/S value is roughly linear with respect to colorant concentration [21].

2.5. FTIR Analysis

To provide insight into the chemical changes occurring over the entire dyeing process, FTIR analysis was carried out on the para-aramid samples at the different stages of treatments. Attenuated total reflection Fourier transform infrared spectroscopy (ATR–FTIR) spectra were recorded using an FTIR spectrometer (Excalibur 3100, Varian Inc., Palo Alto, CA, USA) equipped with an overhead attenuated total reflection (ATR) accessory with germanium crystal (UMA 400, Varian Inc.) and a liquid nitrogen cooled mercury cadmium telluride detector. A sample was placed on a potassium bromide (KBr) plate and pressed under the germanium crystal for scanning. Each spectrum was collected within the mid-IR region from 50–4000 cm^{-1} at a resolution of 4 cm^{-1} after averaging 128 scans. The ATR spectra of samples were presented in absorbance units after taking into account the background spectrum acquired using a blank KBr plate. Between successive measurements, the germanium ATR crystal was carefully cleaned with ethanol, rinsed with distilled water, and dried to prevent cross contamination. All the spectra were ATR-corrected using Varian Resolutions Pro software (Varian Inc., Palo Alto, CA, USA).

2.6. SEM, Extraction Test, Tensile Strength

Scanning Electron Micrographs (SEM) of original para-aramid fibers and dyed samples were acquired on a Zeiss Auriga scanning electron microscope (Carl Zeiss SMT Inc., Oberkochen, Germany).

Complementary to the FTIR analysis, an extraction test was conducted on a set of samples dyed with AA/Soy pretreatment followed by NTP treatment, to investigate how the polymerized AA/Soy binds with the dye and para-aramids. Each sample was extracted with 40 mL of either hexane, ethanol, or 2:1 (v/v) chloroform:methanol of varying polarities at room temperature for 15 h. It was assumed that dye molecules covalently bonded to the fabric and highly polymerized AA/Soy network could not be extracted. The K/S value of each sample before and after the extraction was compared to evaluate the extent of the extraction that reduced the color strength. Furthermore, we replicated a reported novel method [22] for the fabrication of colored materials with significantly reduced dye leaching through covalent immobilization of the desired dye using plasma-generated surface radicals; this plasma dye coating procedure immobilizes a pre-adsorbed layer of a dye functionalized with a radical sensitive group on the surface through radical addition caused by a short NTP treatment. We need to point out that this study demonstrated successful dyeing of some hard-to-dye materials such as inert plastics of polyethylene and polytetrafluoroethylene but dyeing para-aramids was not attempted. We followed the same procedure of dyeing described in this reference [22] using a Disperse Red 1 acrylate dye (CAS Number 13695-46-0, Sigma-Aldrich, St. Louis, MO, USA), which is the only one used in this study that is commercially available, to dye our para-aramids. However, because our NTP source is different from that used in the referenced study, we optimized the NTP treatment time and found that a 30 s NTP time resulted in the highest K/S value (longer time would degrade the dye). Para-aramid samples dyed with the Disperse Red 1 acrylate were subjected to the same extraction procedure for comparison.

To evaluate if our pretreatment and dyeing process would affect the strength of the para-aramids, tensile testing of the para-aramid yarns was performed using a TA.XT plus texture analyzer (Stable Micro Systems, Godalming, UK), following ASTM standard [23]. Yarns of 300 Denier continuous filaments were subjected to the same pretreatment and dyeing process developed in this study, and then placed in a sealed bag to be conditioned for three days together with undyed yarns for comparison. The test was completed in triplicates and peak load at breakpoint was recorded as an indicator of the tensile strength.

3. Results and Discussion

3.1. Analysis of Dyeing Experiments

We designed Experiment A to investigate the impact of AESO concentration, the use of benzyl alcohol (Benzyl-OH) as a swelling agent, and NaCl as an electrolyte on dyeing. Furthermore, we tested the NTP treatment prior to or post AESO pretreatment in order to understand the function of NTP in improving the dyeing color strength. The design of this 4-factor, 2-level, full factorial experiment and results are presented in Table 1.

Table 1. Experiment A: factorial design and the results *.

Sample	AESO conc. (wt.%)	Benzyl-OH (wt.%)	NaCl (wt.%)	NTP Order	K/S **	Std **
A1	30	None	None	Prior	4.23	0.023
A2	30	None	None	Post	9.81	0.121
A3	30	None	8	Prior	6.14	0.075
A4	30	None	8	Post	10.53	0.505
A5	30	2	None	Prior	5.96	0.059
A6	30	2	None	Post	8.49	0.283
A7	30	2	8	Prior	7.56	0.013
A8	30	2	8	Post	10.04	0.081
A9	39	None	None	Prior	3.94	0.122
A10	39	None	None	Post	10.58	0.724
A11	39	None	8	Prior	6.48	0.191
A12	39	None	8	Post	9.22	0.318
A13	39	2	None	Prior	5.91	0.324
A14	39	2	None	Post	9.90	0.350
A15	39	2	8	Prior	7.78	0.202
A16	39	2	8	Post	8.79	0.311

* Experiment conducted at soaking time in AESO in ethyl acetate = 1 h, dye conc. = 0.1 wt.%, swelling agent = 2 wt.%, benzyl alcohol, electrolyte = 8 wt.%, NaCl, NTP treatment time = 90 s, dyeing temperature T = 90 °C, dyeing time = 1 h. ** Mean and standard error of three repeated measurements.

For the analysis of variance (ANOVA) of the factorial design, model term selection (including factorial interactions) was first conducted based on the half-normal probability plot, showing the factorial impact in the ascending order of (AESO conc.) < (Benzyl-OH) < (NaCl) < (NTP order). This resulted in an overall significant model (F-test, $p < 0.0001$). T-tests for each model term coefficients showed that all model terms were significant ($p < 0.002$), except for the main effect of (AESO conc.) ($p = 0.89$). This was not surprising because we did not know a suitable AESO concentration in this first experiment and randomly selected the two AESO concentrations. Since all factors were involved in significant interactions, the statistical inferences are based on the interaction plots as shown in Figure 2.

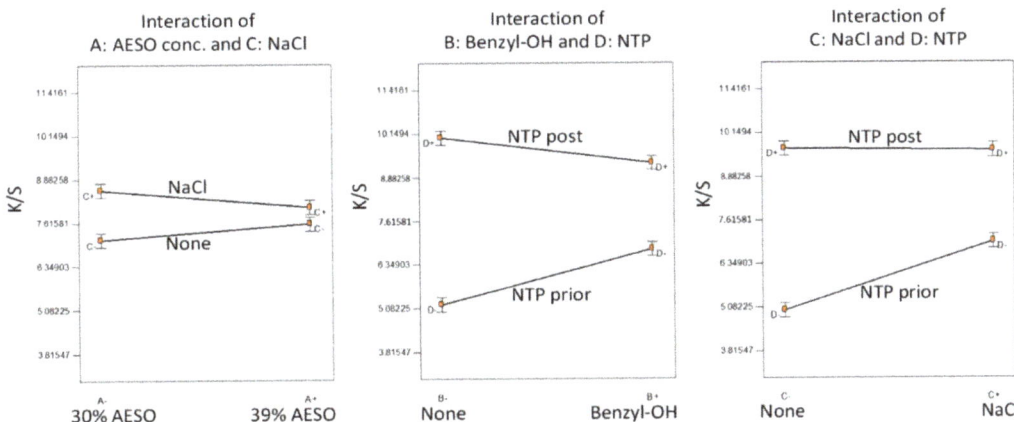

Figure 2. Statistical inferences of Experiment A: significant factorial interactions.

The difference between the two selected AESO concentrations was insignificant. Addition of NaCl in the dye bath slightly improved color strength, apparently when the NTP was applied prior to AESO pretreatment. The order of NTP application played an important role; NTP application post AESO pretreatment greatly increased the dyed color strength, indicating that the chemical reactions induced by NTP on AESO were the key. It appeared that the effect of NTP was confounded by the addition of benzyl alcohol and NaCl in the dye bath, so the two did not show an obvious effect if the fabrics were NTP-treated post AESO pretreatment. A reasonable explanation would be that NTP application prior to AESO pretreatment facilitated the attraction of polar NaCl and benzyl alcohol to the fabrics, while in the case of NTP application post AESO pretreatment of the fabrics was covered by AESO, and NTP was mainly used to induce reactions on the AESO.

Because the interaction between NTP and AESO was important, in Experiment B, we used a central composite design to find optimal AESO concentration and NTP treatment time post AESO pretreatment, without the addition of benzyl alcohol and NaCl in the dye bath. The detailed design and results are presented in Table 2.

Table 2. Experiment B: central composite design and the results *.

Sample	AESO conc. (wt.%)	NTP Time (s)	K/S **	Std **
B1	0.85	75	2.51	0.086
B2	0.85	139	2.54	0.049
B3	5	30	3.63	0.032
B4	5	120	3.98	0.008
B5	15	11	3.80	0.072
B6	15	75	5.25	0.065
B7	15	75	4.65	0.127
B8	15	75	4.73	0.116
B9	15	75	5.25	0.065
B10	15	75	5.54	0.017
B11	15	75	5.77	0.104
B12	15	139	5.45	0.125
B13	25	30	5.05	0.008
B14	25	120	7.16	0.090
B15	29	11	3.55	0.016
B16	29	75	6.84	0.126

* Experiment conducted at soaking time in AESO in ethyl acetate = 15 h, dye conc. = 0.1 wt.%, dyeing temperature T = 90 °C, dyeing time = 1 h. ** Mean and standard error of three repeated measurements.

Stepwise regression resulted in the selection of a reduced quadratic model as the best fit (F-test, $p < 0.001$), including only three terms of the main effect of NTP time, main effect of AESO concentration, and the interaction of the two. The quadratic terms of both NTP time and AESO concentration were not significant.

The response surface plot for Experiment B is shown in Figure 3. Increasing both NTP time and AESO concentration increased color strength. At low AESO concentration, increasing NTP time only slightly increased color strength. However, at higher AESO concentration, increasing NTP time prominently improved color strength. This is reasonable because higher concentration of AESO and longer NTP time induced more polymerized AESO on the fabric surface that attracted more dye molecules, again corroborating the importance of AESO pretreatment followed by NTP treatment.

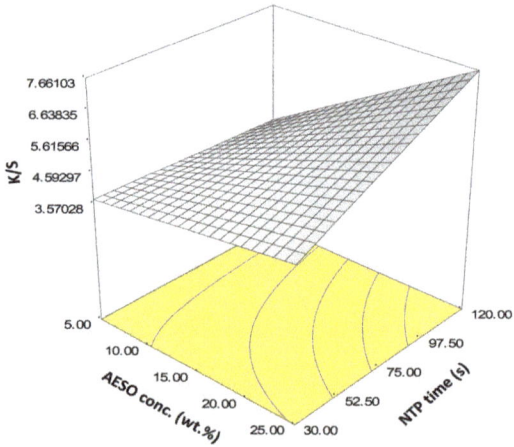

Figure 3. Statistical inference of Experiment B: Response surface of the central composite design.

We also observed that AESO concentration higher than 25 wt.% resulted in an unnecessarily thick coating on the para-aramid fabrics. Therefore, we considered 25 wt.% as an optimal concentration for AESO pretreatment because the highest K/S in Experiment B was achieved at NTP time = 120 s. and AESO conc. = 25 wt.%.

In Experiment C, based on the results of Experiment A and B, we fixed AESO concentration at 25 wt.% and tested if there is any difference between just applying NTP post AESO pretreatment and applying NTP both before and after AESO pretreatment. We also slightly extended the NTP post AESO pretreatment time to 150 s, based on our observation in Figure 3. Experiment C focused on the functions of auxiliary additives of a swelling agent (benzyl alcohol), an electrolyte (NaCl), and a surfactant of Polysorbate 80 (denoted as TWEEN hereafter). The factorial design and results are shown in Table 3.

Table 3. Experiment C: factorial design and the results *.

Sample	Benzyl-OH (wt.%)	NaCl (wt.%)	TWEEN (wt.%)	NTP Order	K/S **	Std **
C1	None	None	None	Both	7.93	0.063
C2	None	None	None	Post	7.94	0.074
C3	None	None	3	Both	7.68	0.019
C4	None	None	3	Post	8.25	0.044
C5	None	8	None	Both	7.47	0.096
C6	None	8	None	Post	7.10	0.164
C7	None	8	3	Both	6.20	0.222

Table 3. Cont.

Sample	Benzyl-OH (wt.%)	NaCl (wt.%)	TWEEN (wt.%)	NTP Order	K/S **	Std **
C8	None	8	3	Post	6.72	0.162
C9	2	None	None	Both	8.46	0.059
C10	2	None	None	Post	7.39	0.092
C11	2	None	3	Both	7.26	0.099
C12	2	None	3	Post	7.58	0.232
C13	2	8	None	Both	8.29	0.171
C14	2	8	None	Post	8.41	0.097
C15	2	8	3	Both	9.23	0.038
C16	2	8	3	Post	6.23	0.491

* Experiment conducted at soaking time in 25 wt.% AESO in ethyl acetate = 15 h, dye conc. = 0.1 wt.%, NTP treatment time = 150 s post AESO soaking (designated Post in Table 3) or 90 s before AESO soaking and 150 s post AESO soaking (designated Both in Table 3), dyeing temperature T = 90 °C, dyeing time = 1 h. ** Mean and standard error of three repeated measurements.

We determined an ANOVA model as the best fit (F-test, $p < 0.0001$) according to the half-normal probability plot, including all the four main factorial effects and only two interaction terms, which were statistically significant based on the t-test of the model coefficients ($p < 0.05$). Statistical inference is presented in Figure 4.

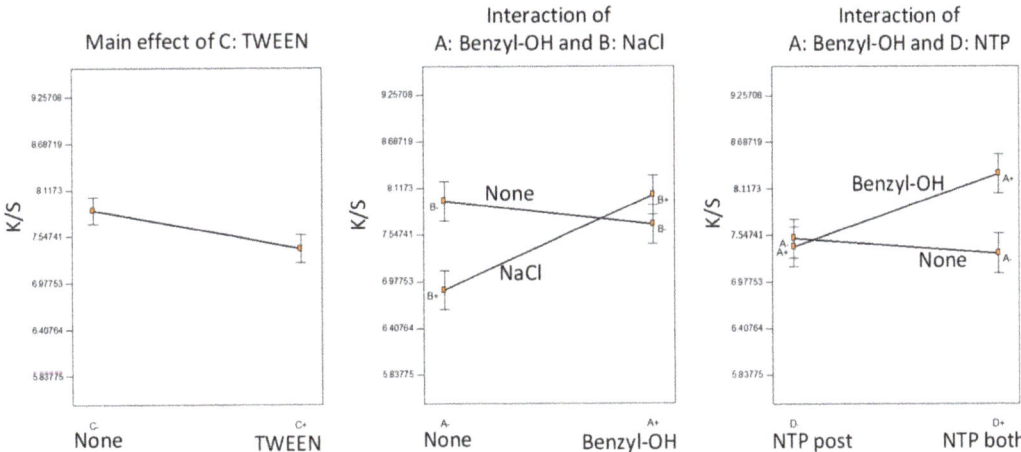

Figure 4. Statistical inference of Experiment C: main effect of a factor not involved in any interaction and significant interactions.

TWEEN was the only factor not involved in any significant interactions; the addition of the surfactant negatively affected dyeing performance, consistent with our previous observations [9,10]. The surfactant, although helping to form a dye dispersion, might also hinder the diffusion of the dye onto the para-aramid fiber fabrics. The positive effect of benzyl alcohol was significant only when the NTP was applied both before and after AESO pretreatment. Furthermore, benzyl alcohol and NaCl were confounding factors. In the absence of benzyl alcohol, the addition of NaCl lowered color strength, but with the presence of benzyl alcohol, the NaCl effect was insignificant. We may speculate that both NaCl and benzyl alcohol participated in the NTP-induced reactions with AESO, leading to different reaction pathways, and therefore, different dyeing effects. Overall, all the samples were dyed to high color strength with small variations because we fixed AESO concentration and NTP treatment time at optimal conditions. The effects of auxiliary additives were limited, compared with the major improvement of color strength caused by AESO pretreatment followed by NTP treatment.

Experiment D was conducted as a starting point to investigate the possibility of replacing AESO with a simple mixture of soybean oil and acrylic acid, because hypothetically a cross-linked network could be formed with the application of NTP to help dyeing. However, longer NTP treatment time is needed based on our preliminary trials. Furthermore, because one soybean oil molecule has an average of 4.6 double bonds and NTP could trigger self-polymerization of soybean oil [24], we tested two molar ratios of acrylic acid to soybean oil (denoted as AA/Soy hereafter) of 1.6 and 4. We also tested if the application of NTP again after dyeing would help fixing the dye. The design and results of this 3-factor, 2-level factorial experiment are shown in Table 4.

Table 4. Experiment D: factorial design and the results *.

Sample	AA/Soy	NTP Time (s)	Post-Dye NTP Time (s)	K/S **	Std **
D1	1.6	240	0	7.67	0.087
D2	1.6	240	30	3.99	0.054
D3	1.6	360	0	9.6	0.175
D4	1.6	360	30	6.85	0.012
D5	4	240	0	8.84	0.056
D6	4	240	30	7.66	0.039
D7	4	360	0	9.59	0.113
D8	4	360	30	8.26	0.023

* Experiment conducted at soaking time in AA/Soy mixture = 15 h, dye conc. = 0.1 wt.%, dyeing temperature T = 90 °C, dyeing time = 1 h. ** Mean and standard error of three repeated measurements.

Based on the half-normal probability plot, we selected an ANOVA model (F-test, $p < 0.0001$), which included all three main factors and two interaction terms (t-test, $p < 0.001$). Because all the factors were involved in the interactions, the statistical inference was based on the significant interaction plots as shown in Figure 5.

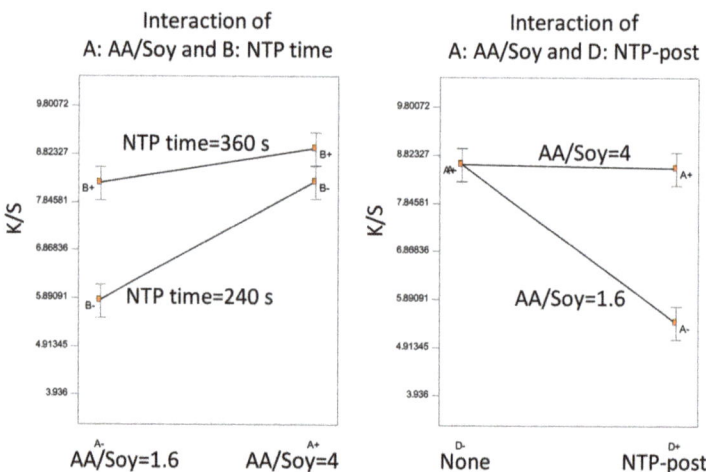

Figure 5. Statistical inference of Experiment D: significant factorial interactions.

Overall, the color strength achieved in Experiment D is comparable with that in Experiment C using AESO, showing the potential of using AA/Soy to replace AESO. Generally, longer NTP treatment time and higher AA/Soy resulted in higher K/S values. It is interesting to observe that at AA/Soy = 1.6, a 30 s NTP treatment after dyeing greatly decreased color strength, whereas at AA/Soy = 4 there was no change. A plausible explanation would be that NTP could quickly degrade the dye molecules [10]; however, at higher AA/Soy, more chemical functionalities could absorb most NTP energy and reactive species, thus protecting the dye from degradation.

In order to evaluate if the use of auxiliary agents would help improve dyeing with the AA/Soy pretreatment, Experiment E focused on the effects of NaCl and benzyl alcohol. Furthermore, because we observed unevenly dyed samples in Experiment D due to inadequate mixing of the soybean oil and acrylic acid, the AA/Soy well dissolved in ethyl acetate (50 wt.% of AA/Soy in ethyl acetate) was used as the pretreatment solution. The design and results of this 4-factor full factorial experiment are presented in Table 5.

Table 5. Experiment E: factorial design and the results *.

Sample	AA/Soy	NTP Time (s)	NaCl (wt.%)	Benzyl-OH (wt.%)	K/S	Std **
E1	1.6	240	None	None	6.18	0.321
E2	1.6	240	None	2	4.6	0.059
E3	4	240	None	None	10.21	0.327
E4	4	240	None	2	9.82	0.589
E5	1.6	240	5	None	8.93	0.42
E6	1.6	240	5	2	7.12	0.034
E7	4	240	5	None	8.69	0.037
E8	4	240	5	2	6.51	0.056
E9	1.6	360	None	None	6.76	0.049
E10	1.6	360	None	2	7.44	0.655
E11	4	360	None	None	7.92	0.051
E12	4	360	None	2	6.74	0.017
E13	1.6	360	5	None	9.01	0.032
E14	1.6	360	5	2	7.18	0.048
E15	4	360	5	None	8.41	0.066
E16	4	360	5	2	7.36	0.031

* Experiment conducted at soaking time in AA/Soy in ethyl acetate = 15 h, dye conc. = 0.1 wt.%, swelling agent = 2 wt.%, benzyl alcohol, electrolyte = 5 wt.%, NaCl, dyeing temperature T = 90 °C, dyeing time = 1 h.
** Mean and standard error of three repeated measurements.

Based on the half-normal probability plot, we selected the best fit of ANOVA model (F-test, $p < 0.0001$). The statistical inference is shown in Figure 6. Only two interaction terms were significant (t-test, $p < 0.01$). However, the main effect of NTP treatment time was not significant (t-test, $p = 0.53$) due to the confounding effect of NaCl.

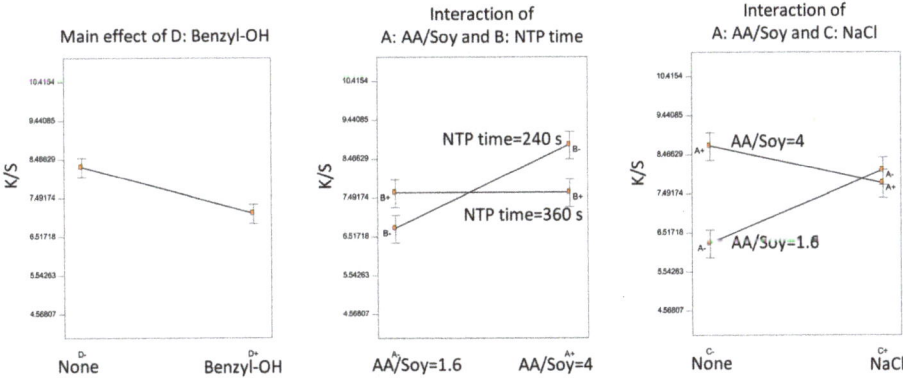

Figure 6. Statistical inference of Experiment E: main effect of a factor not involved in any interaction and significant interactions.

The swelling agent of benzyl alcohol was the only significant main factor not involved in any interactions; it negatively affected the dyeing color strength. In comparison with Experiment D, Experiment E showed similar interactions between the NTP treatment time and the AA/Soy ratio. Specifically, fixing the NTP time = 360 s, increasing AA/Soy from 1.6 to 4, did not improve color strength, while fixing NTP time = 240 s, increasing AA/Soy, significantly increased color strength.

A confounding effect of NaCl was observed. Without NaCl, higher AA/Soy yielded higher color strength; with the addition of NaCl in the dye bath, AA/Soy = 4 and AA/Soy = 1.6 resulted in statistically the same K/S. A plausible explanation would be that both NTP and NaCl would induce reactions in AA/Soy (as revealed in our FTIR analysis in Section 3.2), and the higher dosage of both would result in overreactions that decreased surface functionality for binding with the dye molecules.

Because of the kinetic nature of the reactions catalyzed by NTP and influenced by the presence of NaCl, we speculated that there exists an optimal combination of NTP treatment time, AA/Soy, and NaCl concentration. Therefore, in Experiment F, we utilized a Box–Behnken response surface design in an effort to find this optimal combination. The detailed experimental design and results are presented in Table 6.

Table 6. Experiment F: Box–Behnken design and the results *.

Sample	AA/Soy	NTP Time (s)	NaCl (wt.%)	K/S	Std **
F1	1.5	180	5	10.59	0.019
F2	1.5	270	2	5.56	0.209
F3	1.5	270	8	7.44	0.573
F4	1.5	360	5	10.50	0.387
F5	3.25	180	2	7.96	0.190
F6	3.25	180	8	9.61	0.339
F7	3.25	270	5	13.18	1.157
F8	3.25	270	5	8.94	0.052
F9	3.25	270	5	13.37	0.220
F10	3.25	270	5	12.30	0.259
F11	3.25	270	5	12.16	1.564
F12	3.25	360	2	10.97	0.126
F13	3.25	360	8	10.72	0.063
F14	5	180	5	9.62	0.044
F15	5	270	2	10.94	0.183
F16	5	270	8	9.84	0.125
F17	5	360	5	10.86	0.756

* Experiment conducted at soaking time in AA/Soy (50 wt.%) in ethyl acetate = 15 h, dye conc. = 0.1 wt.%, dyeing temperature T = 90 °C, dyeing time = 1 h. ** Mean and standard error of three repeated measurements.

A reduced quadratic model of response surface was selected as the best fit (F-test, $p < 0.05$) after performing stepwise regression, including the three main factor effects and the quadratic terms of only AA/Soy and NaCl concentration. Representative response surfaces are presented in Figure 7.

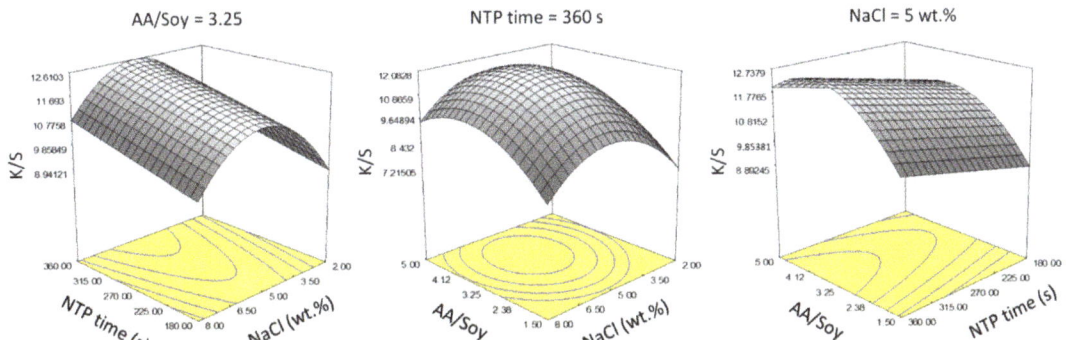

Figure 7. Statistical inference of Experiment F: Response surface of the Box–Behnken design.

Within the selected ranges for the main factors, increasing NTP treatment time linearly increased color strength, so the longest NTP time = 360 s gave the best results. However, there existed an optimal point for AA/Soy and NaCl concentration. Based on the above plots, we determined an optimal condition as NTP time = 360 s, NaCl = 5% wt.%, and AA/Soy = 3.25.

Experiment G was conducted at the determined optimal conditions, and we varied the pretreatment soaking time for practical applications, making it a one-factor factorial design with three replicates. Furthermore, all the samples in Experiment G were washed a second time with detergent by following the same protocol described in Section 2.3, and the change of K/S values (denoted as ΔK/S) was used to evaluate the colorfastness against wash. The results are presented in Figure 8, showing each data point, the means, and the least significant difference (LSD) lines.

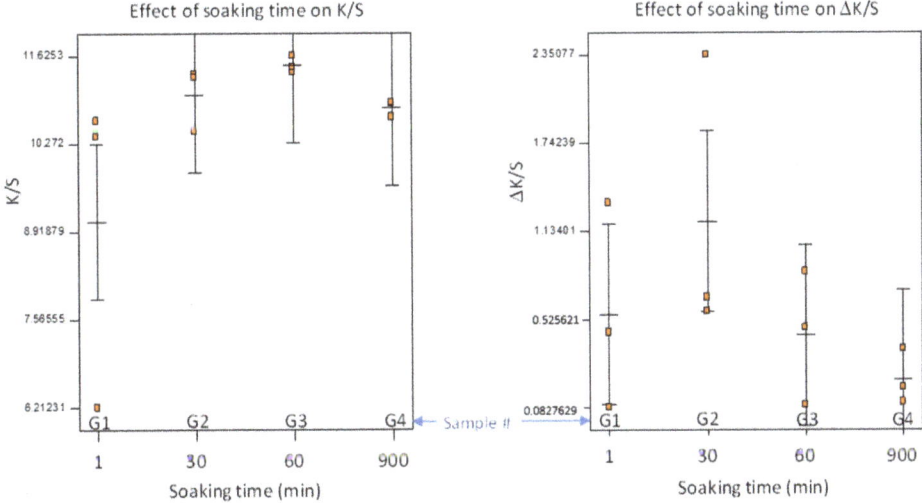

Figure 8. Effect of pretreatment soaking time on color strength and colorfastness (ΔK/S is defined as subtracting K/S after 2nd detergent wash from that after 1st detergent wash).

ANOVA for the one-factor model did not detect any significant difference among the four pretreatment soaking times (F-test, $p > 0.2$), because of a large variation of K/S when the soaking time was short. There was an outlier with low K/S = 6.2, occurring with 1 min of soaking time. Generally, the soaking time appeared to affect the consistency of color strength. The ΔK/S indicated slight reduction in color strength after second detergent wash. Similarly, shorter soaking time resulted in larger variation of ΔK/S. It is recommended that a soaking time ≥60 min be used, or we need to develop a better method to evenly distribute AA/Soy into the fabrics in a shorter time (e.g., continuous soaking and pressing with rollers).

To provide a visual demonstration comparing the dyeing strengths of different treatments, images of selected samples acquired using a flatbed scanner are presented in the Supplementary File (Figure S1).

3.2. FTIR Analysis

To understand chemical changes during the pretreatment and dyeing processes, ATR–FTIR spectroscopy was performed at different stages of the processes. As a foundation for interpreting the FTIR spectra, Table S1 in the Supplementary File summarizes major FTIR bands of AESO, soybean oil, acrylic acid, and para-aramids.

Figure 9 presents ATR–FTIR spectra of materials at different stages of our processes using AESO for pretreatment. The AESO signal is dominated by the major peaks of soybean oil backbone (line I). A strong peak for ester carbonyl groups should appear at 1742 cm^{-1}; however, due to the formation of acrylate esters, a peak should appear at nearby 1723 cm^{-1} [25]. Therefore, the peak with a deformed shape around 1737 cm^{-1} should be the overlapping of 1742 and 1723 cm^{-1}. Compared with the dominant signal from soybean oil backbone, characteristics of AESO/acrylate can be found as small peaks. There are residual epoxides in the AESO that were not fully acrylated as revealed by the oxirane peaks at 1270 and 822 cm^{-1}. Although major acrylate peaks (such as 1638 and 1619 cm^{-1}) overlap with a strong signal from para-aramids, two distinguishable acrylate peaks at 1406 and 810 cm^{-1} can be tracked along the way of our processes to indicate the consumption of the acrylates.

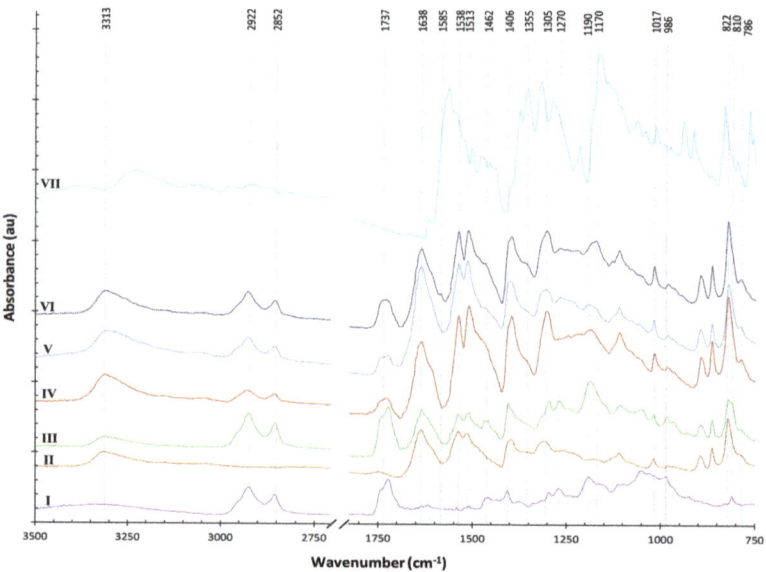

Figure 9. FTIR spectra at different stages of materials of (I) AESO; (II) Para-aramids; (III) Para-aramid after soaking in AESO; (IV) Para-aramid after soaking in AESO and subsequent NTP treatment; (V) dyed sample A3 pretreated with AESO without subsequent NTP treatment; (VI) dyed sample C9 pretreated with AESO with subsequent NTP treatment; (VII) Victoria Blue R.

For para-aramids, the bands shown in Table S1 and correspondingly in Figure 9 (line II) can be considered as an internal standard [26]. The FTIR spectrum of AESO-soaked para-aramids apparently show the addition of signals from both AESO and para-aramids (line III). After subsequent NTP treatment, changes can be seen (line IV): the acrylate peaks at 1406 and 810 cm^{-1} disappeared or significantly decreased. Although this observation was somewhat obscured by a nearby peak of para-aramids, the shapes of the two peaks more closely resemble those of the para-aramid peaks nearby, indicating the consumption of the acrylate functional groups. A sobservation can be found around 1737 cm^{-1}, showing the consumption of 1723 cm^{-1} acrylate esters. The oxirane peak at 1270 cm^{-1} also disappeared (line IV). However, the signal of vinyl functional group at 986 cm^{-1} remained, albeit decreased, indicating that the AESO was not fully polymerized. The 1190 cm^{-1} peak of C–O stretching in the esters of soybean oil backbone also remained, revealing that the NTP treatment did not fully break the ester bonds.

Figure 9 also shows the FTIR spectrum of a dyed sample which was pretreated with AESO without subsequent NTP treatment (line V, sample A3, K/S = 6.14), and another dyed sample pretreated with AESO with subsequent NTP treatment (line VI, sample C9, K/S = 8.46). Other than differences at a few minor peaks, lines V and VI are very similar. The –C=O carbonyl stretching and C–O stretching in the esters at 1737 cm^{-1} and 1190 cm^{-1}, respectively remained strong in lines V and VI, indicating that the NTP treatment and the dyeing process did not break much of the ester bonds, which remained after dyeing and detergent wash. However, the 1190 cm^{-1} peak in line VI shifted to the right, probably due to the influence of the dye peak at 1170 cm^{-1}, because the sample of line VI was dyed to a higher K/S value. The three spectra (lines IV, V, and VI) all have a new peak emerging at 786 cm^{-1}, which is in the typical region of 1,3-disubstituted or 1,2,3-trisubstituted C–H bending, meaning a possible new attachment of AESO to the para-aromatic rings. Interestingly, this new peak at 786 cm^{-1} could be generated by the NTP treatment or the thermal dyeing process, because the sample of line V was not treated with NTP after AESO soaking. In line VI, the weak absorption band at 1585 cm^{-1} was assigned to aromatic rings conjugated with α-carbonyl group, indicating that some AESO covalently bonded to para-aramid rings after NTP treatment [27].

Additionally, lines IV, V, and VI obviously have stronger FTIR signal because the AESO pretreatment, NTP treatment, and dyeing process changed the reflective index of the treated samples, enabling deeper ATR–FTIR penetration.

The FTIR spectrum of the Victoria Blue R dye is also shown in Figure 9 (line VII). Because of low uptake of the dye by the fabrics and that most of the dye peaks were obscured by the characteristic peaks of the para-aramids and AESO, no significant dye signal was detected in the dyed samples, except for that at 1355 and 1170 cm^{-1} that rendered a small shoulder in the spectra of dyed samples (lines V and VI), especially clear for the sample of line VI, which was treated with NTP and consequently dyed to a higher color strength.

Because the strong signal of para-aramids obscured many meaningful peaks that reveal chemical changes, we used ATR–FTIR to monitor reactions between soybean oil and acrylic acid on KBr plates (instead of on para-aramids) in contrast to those of AESO. It is important to note that the reagents acrylic acid and AESO contain 200 or 4000 ppm of monomethyl ether of hydroquinone (MEHQ), respectively, as an inhibitor of self-polymerization. Figure 10 shows the results; line I is the spectrum of soybean oil; line II is that of acrylic acid; and line III is that of the mixture of acrylic acid and soybean oil in a molar ratio of 4:1, apparently showing the signal addition of soybean oil and acrylic acid.

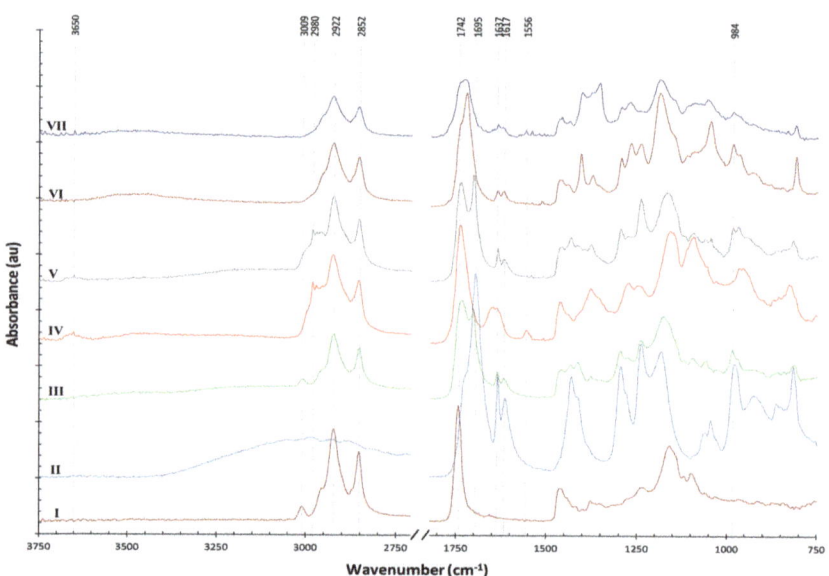

Figure 10. FTIR spectra at different stages of materials on KBr of (I) soybean oil; (II) acrylic acid; (III) mixture of soybean oil and acrylic acid; (IV) mixture of soybean oil and acrylic acid after 6 min NTP treatment; (V) mixture of soybean oil, acrylic acid, and NaCl; (VI) AESO; (VII) AESO after 2.5 min NTP treatment.

After a 6 min NTP treatment of the mixture of soybean oil and acrylic acid, significant changes could be observed (line IV). A small but obvious broad peak appeared around 3650 cm^{-1}, which is assigned to O–H stretching usually from alcohols or phenols. Because the NTP generated in ambient air has ozone, singlet oxygen atoms, and hydrogen peroxide as highly reactive oxidants, the unstable C=C double bonds in soybean oil could be first attacked to form epoxidized soybean oil, which could be followed by ensuing reaction with acrylic acid catalyzed by UV and/or radical species in NTP to form AESO. However, a competitive reaction could also lead to the production of polyols [28], which is evident in line IV, and the polyols could react with acrylic acid to form acrylate polymers [29,30].

After NTP treatment (line IV), the 3009 cm^{-1} peak (=C–H in soybean oil) disappeared and some new peaks appeared between 3009 and 2922 cm^{-1}. The peak at 2980 cm^{-1} can be assigned to C–H (sp^3 stretching), indicating hydrogen abstraction from the –CH$_3$ or –CH$_2$ of the soybean oil. However, if the FTIR spectrum was acquired on para-aramids pretreated with AA/Soy followed by NTP treatment, (see line A in Figure S2 of Supplementary File), this peak at 2980 cm^{-1} was not observed. Instead, a small peak appeared at 786 cm^{-1}, which is the typical region of 1,3-disubstituted or 1,2,3-trisubstituted C–H bending and similar to the case in Figure 9 for samples pretreated with AESO. Therefore, we reasoned that the AESO or AA/Soy network was "anchored" to para-aramids at the C–H point.

The 1695 cm^{-1} peak (C=O stretching) in acrylic acid was significantly consumed with the application of NTP, and a new peak at 1556 cm^{-1} indicated nitro groups caused by the reactive nitrogen and oxygen species (RONS) in NTP; a small peak at the same wavelength can also be observed in NTP-treated AESO (line VII).

Interestingly, by simply adding NaCl to the mixture of soybean oil and acrylic acid, similar changes can be observed around 3650 and 2980 cm^{-1} (line V), indicating that NaCl triggered reactions between the soybean oil and acrylic acid at room temperature, which explains the confounding effect of NaCl we observed in Experiment E, in that the reactions with or without NaCl might lead to different kinetics and pathways. With NTP application on the mixture of soybean oil and acrylic acid (line III vs. VI), the 1637 and 1617 cm^{-1} of

C=C double bonds in the acrylic acid were consumed, indicating that the NTP treatment did induce radical polymerization through the C=C double bonds. A similar observation could be found in NTP-treated AESO (line VI vs. VII). However, for the NTP-treated mixture of soybean oil and acrylic acid, the 984 cm^{-1} peak (=CH$_2$) was almost completely consumed after a 6 min NTP treatment (line IV), whereas the 986 cm^{-1} peak of vinyl functional groups in AESO was still visible (line VII), indicating that a 2.5 min NTP treatment did not totally consume the vinyl groups, and longer NTP time could be applied on the AESO. The FTIR spectra of dyed samples pretreated with AESO or AA/Soy were very similar, except for some small peaks (see line B vs. C in Figure S2 of Supplementary File).

The NTP-induced chemical reactions in AESO and AA/Soy are complex, and the precise details of reaction pathways on the surface of para-aramids deserve further study. Nonetheless, based on the FTIR results, we can propose the following mechanism of improved dyeing with the pretreatment of AESO or AA/Soy. Because of the high crystallinity and inertness of para-aramids, few dye molecules can be directly adsorbed on the para-aramid surface. With the pretreatment of AESO or AA/Soy and subsequent NTP treatment, a cross-linked network (yet flexible due to the preserved long acyl chains) is formed, "anchoring" to the para-aramids with covalent bonds, and also providing functional groups that bind the dye molecules. Although the pretreatment with AESO vs. AA/Soy generated differences in the polymerized network on para-aramids, comparable dyeing effects were achieved, as evident in Section 3.1.

3.3. Supporting Evidence: SEM, Extraction Test, and Tensile Strength

Scanning electron micrographs (SEM) reveal the smooth surface of original para-aramid fibers, and a 4 min NTP treatment did not show any visible changes on the surface (Figure 11), explaining why the NTP treatment alone would not significantly improve dyeing. The samples dyed with AESO or AA/Soy pretreatment have a resin-like coating on the para-aramid fibers, in which dye particles are embedded. It is notable that the coating survived two washes with detergent, indicating a polymerized network. Agglomeration of dye particles can also be observed, especially in the AA/Soy pretreated sample, indicating that our pretreatment procedure needs further improvement in order to evenly distribute the AESO or AA/Soy and deliver the NTP treatment.

After a 15 h extraction with three solvents of different polarity, the color strength of all the tested samples decreased. Because the K/S values were measured at different wavelengths for the blue cationic dye vs. the red disperse dye, we report here the percentage decrease in K/S value after extraction for comparison. For the samples dyed with AA/Soy pretreatment followed by NTP treatment, the K/S value decreased by 20%, 64%, and 63% after the extraction using hexane, ethanol, or 2:1 (v/v) chloroform:methanol, respectively. For the sample dyed with Disperse Red 1 acrylate, the K/S value decreased by 6%, 54%, and 84% after the extraction using hexane, ethanol, or 2:1 (v/v) chloroform:methanol, respectively. This indicates that not all the AA/Soy were polymerized (and the unpolymerized soybean oil could be easily extracted by hexane), or not all the AA/Soy were covalently bonded to the fabric. However, there was evidence that almost all the dye molecules were covalently bonded to the NTP-induced polymerized network. When the solvents were evaporated from the extractable matters and the dry lipid-dye matrices were re-dissolved in hexane, then water was added to study the partition of the dye between the hexane and water, no blue color was observed in the aqueous layer, indicating that this water-soluble dye was chemically linked to the soybean oil network.

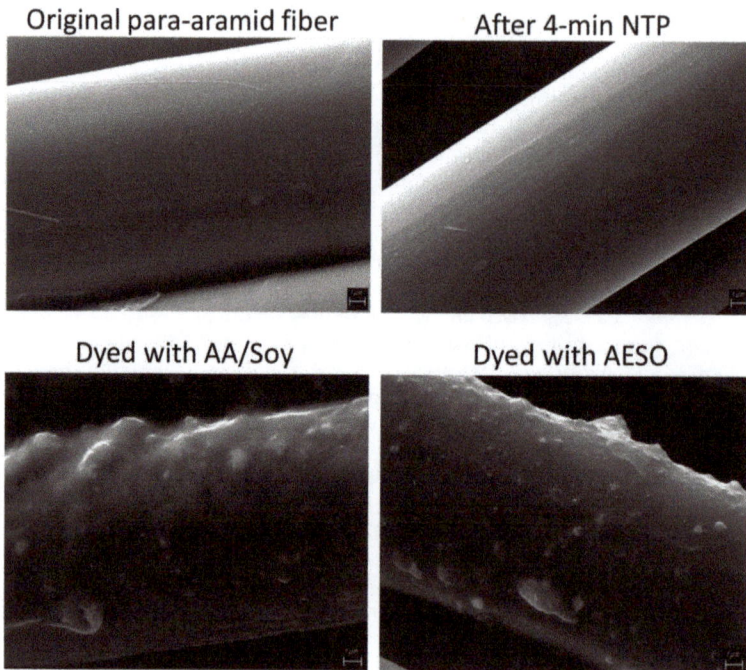

Figure 11. SEM micrograph of samples.

In addition, significant color strength remained after extraction with the three solvents of different polarities, indicating a covalently bonded network formed on the surface of para-aramid fabrics, with one exception of the 84% reduction in K/S value for the sample dyed with Disperse Red 1 acrylate and extracted with chloroform:methanol, which indicates the lack of bonding between the polymerized dye and the para-aramids; the extracted sample exhibited a yellow color close to that of the undyed fabric. In this respect, the method developed in this study performed better in binding with the inert para-aramids. Furthermore, the samples dyed with Disperse Red 1 acrylate became more rigid because of the short chain of the starting monomers, while the samples dyed with AA/Soy remained flexible due to the preserved long chains of soybean oil.

It is worthwhile to mention that in dyeing textiles, dyes are usually fixed on textiles through physical entrapment or the relatively weak hydrogen bonds and van der Waals force of attraction, or stronger ionic bonds in the case of acid dye or basic dye; few dye molecules binded to textiles via covalent bonds, except for reactive dyes with cotton, wool, or polyamides such as nylon [31,32].

Tensile test results show that our developed method of dyeing para-aramids did not reduce the strength of para-aramid fibers measured as breakpoint peak load.

4. Conclusions

We demonstrated that deriving more functionalities of soybean oil on the surface of para-aramids further improved the dyeing color strength and colorfastness. Compared with the pretreatment using soybean oil that obtained the highest K/S value up to 3.89, pretreatment with either AESO or AA/Soy achieved K/S values higher than nine in this study. Importantly, we demonstrated through sequential experimentation that a simple mixture of AA/Soy could replace the pricy AESO to render comparable dyeing performance. NTP treatment after the AESO or AA/Soy pretreatment was essential for our developed method, because the NTP not only induced the formation of a polymerized network on the surface of para-aramids that bonded with dye molecules, but also generated covalent bonds anchoring the polymerized network to the para-aramids, which is difficult given the high crystallinity and chemical inertness of the para-aramids.

Among the auxiliary additives tested in this study, Polysorbate 80 (TWEEN) as a surfactant negatively affected the dyeing, and the effect of benzyl alcohol as a swelling agent was minimal. However, NaCl as an electrolyte showed positive effect. Therefore, we recommend an optimal formulation and condition as NTP time = 360 s (specific for the NTP source used in this study), NaCl = 5% wt.%, and AA/Soy = 3.25 for dyeing para-aramids without other chemical additives. The dyeing method developed in this study did not sacrifice the strength of para-aramids, showing the potential for this method to replace hazardous chemicals currently used in dyeing practices with renewable materials and environmentally friendly technologies, while achieving improved dyeing of para-aramid textiles for new applications.

Supplementary Materials: The following supporting information can be downloaded at: https://www.mdpi.com/article/10.3390/polym14030628/s1, Table S1: Characteristic FTIR bands of materials and the references; Figure S1: Scanned images of dyed samples in contrast with original undyed fabric (labels correspond to the sample numbers in Tables 3–5 and Figure 8; G4 (W2) indicates sample G4 after 2nd detergent wash); Figure S2: FTIR spectra of (A) para-aramid sample after soaking in AA/Soy and subsequent NTP treatment; (B) dyed para-aramid sample pretreated with AA/Soy with subsequent NTP treatment; (C) dyed para-aramid sample pretreated with AESO with subsequent NTP treatment.

Author Contributions: Conceptualization, C.J.D. and X.P.Y.; methodology, X.P.Y. and C.J.D.; validation, X.P.Y., T.W. and C.M.; formal analysis, C.J.D., C.M. and X.P.Y.; investigation, C.M., T.W. and X.P.Y.; resources, C.J.D. and X.P.Y.; data curation, C.M. and X.P.Y.; writing—original draft preparation, X.P.Y.; writing—review and editing, X.P.Y., T.W. and C.J.D.; visualization, C.M. and X.P.Y.; supervision, X.P.Y. and T.W.; project administration, C.J.D. and X.P.Y.; funding acquisition, C.J.D. and X.P.Y. All authors have read and agreed to the published version of the manuscript.

Funding: This work was partially supported by the Department of Agriculture HATCH project No. TEN00521.

Institutional Review Board Statement: Not applicable.

Informed Consent Statement: Not applicable.

Data Availability Statement: The data presented in this study are available on request from the corresponding author.

Acknowledgments: We thank the support of the U.S. Department of Agriculture HATCH project No. TEN00521.

Conflicts of Interest: The authors declare no conflict of interest.

References

1. Sabir, T. 2—Fibers Used for High-Performance Apparel. In *High-Performance Apparel*; McLoughlin, J., Sabir, T., Eds.; Woodhead Publishing: Sawston, UK, 2018; pp. 7–32.
2. Gong, R.H.; Chen, X. 3—Technical Yarns. In *Handbook of Technical Textiles*, 2nd ed.; Horrocks, A.R., Anand, S.C., Eds.; Woodhead Publishing: Sawston, UK, 2016; pp. 43–62.
3. Roenbeck, M.R.; Cline, J.; Wu, V.; Afshari, M.; Kellner, S.; Martin, P.; Londono, J.D.; Clinger, L.E.; Reichert, D.; Lustig, S.R.; et al. Structure–property relationships of aramid fibers via X-ray scattering and atomic force microscopy. *J. Mater. Sci.* **2019**, *54*, 6668–6683. [CrossRef]
4. Roenbeck, M.R.; Sandoz-Rosado, E.J.; Cline, J.; Wu, V.; Moy, P.; Afshari, M.; Reichert, D.; Lustig, S.R.; Strawhecker, K.E. Probing the internal structures of Kevlar® fibers and their impacts on mechanical performance. *Polymer* **2017**, *128*, 200–210. [CrossRef]
5. Sun, Y.; Liang, Q.; Chi, H.; Zhang, Y.; Shi, Y.; Fang, D.; Li, F. The application of gas plasma technologies in surface modification of aramid fiber. *Fibers Polym.* **2014**, *15*, 1–7. [CrossRef]
6. Xi, M.; Li, Y.-L.; Shang, S.-Y.; Li, D.-H.; Yin, Y.-X.; Dai, X.-Y. Surface modification of aramid fiber by air DBD plasma at atmospheric pressure with continuous on-line processing. *Surf. Coat. Technol.* **2008**, *202*, 6029–6033. [CrossRef]

7. Su, M.; Gu, A.; Liang, G.; Yuan, L. The effect of oxygen-plasma treatment on Kevlar fibers and the properties of Kevlar fibers/bismaleimide composites. *Appl. Surf. Sci.* **2011**, *257*, 3158–3167. [CrossRef]
8. Kaul, B.L. *Synthetic Fibre Dyeing*; Hawkyard, C., Ed.; Society of Dyers and Colourists: Bradford, UK, 2004; p. 230.
9. Morris, M.; Ye, X.P.; Doona, C.J. Soybean Oil and Nonthermal Plasma Pretreatment to Dye Para-Aramid Woven Fabrics with Disperse Dye Using a Glycerol-Based Dye Bath. *J. Am. Oil Chem. Soc.* **2021**, *98*, 463–473. [CrossRef]
10. Morris, M.; Ye, X.P.; Doona, C.J. Dyeing Para-Aramid Textiles Pretreated with Soybean Oil and Nonthermal Plasma Using Cationic Dye. *Polymers* **2021**, *13*, 1492. [CrossRef]
11. Omonov, T.S.; Curtis, J.M. 7—Plant Oil-Based Epoxy Intermediates for Polymers. In *Bio-Based Plant Oil Polymers and Composites*; Madbouly, S.A., Zhang, C., Kessler, M.R., Eds.; William Andrew Publishing: New York, NY, USA, 2016; pp. 99–125.
12. Behera, D.; Banthia, A.K. Synthesis, characterization, and kinetics study of thermal decomposition of epoxidized soybean oil acrylate. *J. Appl. Polym. Sci.* **2008**, *109*, 2583–2590. [CrossRef]
13. Zou, B.; Ren, S.; Ye, X.P. Glycerol Dehydration to Acrolein Catalyzed by ZSM-5 Zeolite in Supercritical Carbon Dioxide Medium. *ChemSusChem* **2016**, *9*, 3268–3271. [CrossRef]
14. Liu, L.; Ye, X.P.; Bozell, J.J. A Comparative Review of Petroleum-Based and Bio-Based Acrolein Production. *ChemSusChem* **2012**, *5*, 1162–1180. [CrossRef]
15. Ye, X.P.; Ren, S. 3—Value-Added Chemicals from Glycerol. In *Soy-Based Chemicals and Materials*; Brentin, R., Ed.; American Chemical Society: Washington, DC, USA, 2014; pp. 43–80.
16. Baliarsingh, S.; Jena, J.; Das, T.; Das, N.B. Role of cationic and anionic surfactants in textile dyeing with natural dyes extracted from waste plant materials and their potential antimicrobial properties. *Ind. Crops Prod.* **2013**, *50*, 618–624. [CrossRef]
17. Arif, M.F.; Butt, M.T.; Khan, M.N.; Ali, M. Effect of electrolyte concentration on dyeing process of cotton. *Pak. J. Sci. Ind. Res.* **2007**, *50*, 159–164.
18. Shao, D.; Xu, C.; Wang, H.; Du, J. Enhancing the Dyeability of Polyimide Fibers with the Assistance of Swelling Agents. *Materials* **2019**, *12*, 347. [CrossRef]
19. *Standard ISO 105-C10:2006*; Colour Fastness to Washing with Soap or Soap and Soda. Textiles—Tests for Colour Fastness. ISO: London, UK, 2006.
20. Etters, J.N.; Hurwitz, M.D. Opaque reflectance of translucent fabric. *Text. Chem. Colorist* **1986**, *18*, 19–26.
21. Hunter, R.S.; Harold, R.W. *The Measurement of Appearance*, 2nd ed.; Wiley: New York, NY, USA, 1987.
22. De Smet, L.; Vancoillie, G.; Minshall, P.; Lava, K.; Steyaert, I.; Schoolaert, E.; Van De Walle, E.; Dubruel, P.; De Clerck, K.; Hoogenboom, R. Plasma dye coating as straightforward and widely applicable procedure for dye immobilization on polymeric materials. *Nat. Commun.* **2018**, *9*, 1123. [CrossRef]
23. ASTM. *Standard Test Methods for Tensile Testing of Aramid Yarns*; D7269/D7269M−17; ASTM International: West Conshohocken, PA, USA, 2017.
24. Ximena, V.Y. Characterization and Analysis of High Voltage Atmospheric Cold Plasma Treatment of Soilbean Oil. Ph.D. Thesis, Purdue University, West Lafayette, IN, USA, 2020.
25. Esen, H.; Çayli, G. Epoxidation and polymerization of acrylated castor oil. *Eur. J. Lipid Sci. Technol.* **2016**, *118*, 959–966. [CrossRef]
26. Mukherjee, M.; Kumar, S.; Bose, S.; Das, C.; Kharitonov, A. Study on the Mechanical, Rheological, and Morphological Properties of Short Kevlar™ Fiber/s-PS Composites. *Polym. Plast. Technol. Eng.* **2008**, *47*, 623–629. [CrossRef]
27. Van der Weerd, J.; van Loon, A.; Boon, J.J. FTIR Studies of the Effects of Pigments on the Aging of Oil. *Stud. Conserv.* **2005**, *50*, 3–22. [CrossRef]
28. Santacesaria, E.; Turco, R.; Russo, V.; Tesser, R.; Di Serio, M. Soybean Oil Epoxidation: Kinetics of the Epoxide Ring Opening Reactions. *Processes* **2020**, *8*, 1134. [CrossRef]
29. Cheong, M.; Ooi, T.; Ahmad, S.; Yunus, W.; Kuang, D. Synthesis and Characterization of Palm-Based Resin for UV Coating. *J. Appl. Polym. Sci.* **2008**, *111*, 2353–2361. [CrossRef]
30. Li, Y.; Sun, S. Polyols from epoxidized soybean oil and alpha hydroxyl acids and their adhesion properties from UV polymerization. *Int. J. Adhes. Adhes.* **2015**, *63*, 1–8. [CrossRef]
31. Nicolai, S.; Tralau, T.; Luch, A.; Pirow, R. A scientific review of colorful textiles. *J. Consum. Prot. Food Saf.* **2021**, *16*, 5–17. [CrossRef]
32. Fleischmann, C.; Lievenbrück, M.; Ritter, H. Polymers and Dyes: Developments and Applications. *Polymers* **2015**, *7*, 717–746. [CrossRef]

Article

Simultaneous Sonochemical Coloration and Antibacterial Functionalization of Leather with Selenium Nanoparticles (SeNPs)

Tarek Abou Elmaaty [1,2,*], Khaled Sayed-Ahmed [3], Radwan Mohamed Ali [4], Kholoud El-Khodary [2] and Shereen A. Abdeldayem [2]

1. Department of Material Art, Galala University, Galala 43713, Egypt
2. Department of Textile Printing, Dyeing and Finishing, Faculty of Applied Arts, Damietta University, Damietta 34512, Egypt; khloud.el.khodary@gmail.com (K.E.-K.); shereen.abdeldayem.82@gmail.com (S.A.A.)
3. Department of Agricultural Chemistry, Faculty of Agriculture, Damietta University, Damietta 34512, Egypt; drkhaled_1@du.edu.eg
4. Department of Biochemistry, Faculty of Agriculture, Al-Azhar University, Cairo 11651, Egypt; yaseenjjar@gmail.com
* Correspondence: tasaid@gu.edu.eg

Abstract: The development of antibacterial coatings for footwear components is of great interest both from an industry and consumer point of view. In this work, the leather material was developed taking advantage of the intrinsic antibacterial activity and coloring ability of selenium nanoparticles (SeNPs). The SeNPs were synthesized and implemented into the leather surface by using ultrasonic techniques to obtain simultaneous coloring and functionalization. The formation of SeNPs in the solutions was evaluated using UV/Vis spectroscopy and the morphology of the NPs was determined by transmission electron microscopy (TEM). The treated leather material (leather/SeNPs) was characterized by scanning electron microscopy (SEM) and energy dispersive X-ray spectroscopy (EDX). The effects of SeNPs on the coloration and antibacterial properties of the leather material were evaluated. The results revealed that the NPs were mostly spherical in shape, regularly distributed, and closely anchored to the leather surface. The particle size distribution of SeNPs at concentrations of 25 mM and 50 mM was in the range of 36–77 nm and 41–149 nm, respectively. It was observed that leather/SeNPs exhibited a higher depth of shade compared to untreated ones, as well as excellent fastness properties. The results showed that leather/SeNPs can significantly enhance the antibacterial activity against model of bacteria, including Gram-positive bacteria (*Bacillus cereus*) and Gram negative bacteria (*Pseudomonas aeruginosa*, *Salmonella typhi* and *Escherichia coli*). Moreover, the resulting leather exhibited low cytotoxicity against HFB4 cell lines. This achievement should be quite appealing to the footwear industry as a way to prevent the spread of bacterial infection promoted by humidity, poor breathability and temperature which promote the expansion of the microflora of the skin.

Keywords: selenium nanoparticles; leather; sonochemical; antibacterial; toxicity; footwear

1. Introduction

Leather is a durable and flexible material made by tanning animal rawhide and skin, most commonly cattle hide [1]. It has unique characteristics, including high tensile strength, elasticity, tear resistance, high porosity, and air/water permeability [2,3]. In addition, collagen is the main component of natural leather and amino acids are the main units of collagen. It contains functional groups such as $–NH_4^+$, $–COO^-$ as well as OH groups [4]. Collagen of leather can provide ideal conditions such as moisture, temperature, oxygen, and nutrients required for the rapid growth of bacteria and fungi [5].

Leathers have undergone many developments concerning their color and acquired functionalization such as antibacterial, self-cleaning, and UV protection [6,7]. In addition, diverse dyes and dyeing methods have been utilized for sustainable development in the leather industry [8]. The most important of these developments is the use of nanotechnology, which has an eco-friendly effect on the environment [9,10].

Nowadays, coloration and functionalization of leather surfaces with organic nanoparticles have received considerable attention [7,11–16]. The multifunctional features of leather, such as antimicrobial and UV radiation protection, as well as self-cleaning capabilities, have improved leather's adaptability [1,6,9,17–19].

Recently, many studies have reported the use of AgNPs to functionalize leather surfaces due to their antibacterial properties [5,20–23]. In addition, silicon nanoparticles were used with polyurethanes and polyacrylates to produce leather with high water vapor permeability, good mechanical properties as well as good thermal properties [24,25]. Additionally, Ag-TiO$_2$ nanoparticles were deposited on the leather surface to obtain antimicrobial properties and improve photocatalytic activity [26,27]. As well, TiO$_2$-SiO$_2$ nanocomposite was used to enhance the performance such as colorfastness and adhesion [28].

From what was mentioned previously, we found that the multifunctional properties of leathers improved by different kinds of nanoparticles, which were investigated to evaluate their antimicrobial, mechanical, and thermal properties, as well as UV protection and photocatalytic activity. However, the use of NPs as colorants and finishing agents in leather processing has not been sufficiently investigated. Moreover, leather finishing with NPs requires multiple steps, a long reaction time, and the use of more hazardous chemical agents. As a result, there is a great demand for a simple and environmentally safe process for the incorporation of nanoparticles into leather surfaces without the use of hazardous chemical ingredients. One of the most effective nanoparticles in this context is selenium nanoparticles (SeNPs) which have excellent coloring capability [29–32], antimicrobial, antitumor, antioxidant, and antibiofilm properties [33–37]. Moreover, SeNPs exhibit low cytotoxicity [29,38].

In this work we have developed a simple technique to obtain simultaneous coloration and functionalization of leather via the incorporation of SeNPs into the leather under study utilizing an ultrasonic technique. The SeNPs was synthesized through a green process and used as a colorant without any harmful additives, achieving outstanding color fastness and antibacterial properties.

To the best of our knowledge, the utilization of SeNPs for coloration and functionalization of leather has not been reported elsewhere.

2. Materials and Methods

2.1. Materials

Salted wet cowhide leathers were obtained from the Timex Tannery Company, Cairo, Egypt, and then tanned, as reported by Ali et al. [39]. As for chemicals; sodium hydrogen selenite, ascorbic acid, and polyvinylpyrrolidone (PVP) were purchased from Loba Chemie, Mumbai, India. Other chemicals were of commercial grade.

2.2. Green Synthesis of SeNPs

SeNPs were prepared via redox reaction as reported by Abou Elmaaty et al. [32] with an improved modification. Sodium hydrogen selenite was used at different concentrations (50 mM and 100 mM) as a precursor for SeNPs. Polyvinylpyrrolidone (PVP) was dissolved in sodium hydrogen selenite solution at a concentration of 12 g/100 mL to maintain the nanoparticles stability. Then, ascorbic acid at varied concentrations of 100 mM and 200 mM was mixed with this mixture at a volume ratio of 1:1 and molar ratio of 2:1 (ascorbic acid: NaHSeO$_3$). The color change from colorless to dark orange indicated the formation of SeNPs.

2.3. Implementation of Selenium Nanoparticles (SeNPs) into Leather

Before treatment, the leather samples were scoured with nonionic detergent (3%) on a weight of leather at 50 °C for 15 min at a liquor ratio of 1:50 to remove impurities. Then, the leather sample was rinsed with distilled water, followed by treatment into a solution of SeNPs at a liquor ratio (LR) of 1:50 by ultrasonic water bath. The treatment was carried out at varied temperatures (at room, 40, 50, 60, 65 and 70 °C), different periods (30, 60, 90 and 120 min) as well as different *pH* values (3,4,5,6,7,8) and two concentrations of SeNPs of (25 and 50) mmol/l. Subsequently, the leather sample was removed, rinsed with nonionic detergent (2%) on weight of leather at 40 °C for 15 min at a liquor ratio of 1:50, rinsed with distilled water, and dried in an oven at 50 °C for 25 min. Finally, the leather samples were placed in a desiccator to remove traces of water [4].

2.4. Transmission Electron Microscopy (TEM) Analysis

SeNPs characteristics included morphology and size were characterized by transmission electron microscope (JEOL, JEM 2100F) at 200 kV. A drop of the NPs colloidal solution was loaded onto a 400-mesh copper grid with an amorphous carbon film, and water was evaporated in air at room temperature.

2.5. Leather Characterization

2.5.1. SEM and EDX Analysis

The surface morphology of the leather samples, including blank samples and those treated with SeNPs, was observed using a scanning electron microscope (JEOL JSM-6510LB, Tokyo, Japan) with attached energy dispersive X-ray spectrum (EDX) unit.

2.5.2. Raman Spectroscopy Analysis

The types of bonds found in the blank or treated leather were detected using a confocal Raman microscope (Jasco NRS-4500, Tokyo, Japan) at the range of 200–1600 cm^{-1}. For both Raman data acquisition and processing, Jasco spectroscopy suite software was used.

2.5.3. Colorimetric Study

The color uptake, expressed as the color strength (*K/S*) of blank leather and leather/SeNPs, was determined using a spectrophotometer (CM3600A; Konica Minolta, Japan). *K/S* values were evaluated at the wavelength of maximum absorption (λ_{max}) of the color's reflectance curve at 360 nm. The total color difference (ΔE) was represented in terms of CIE LAB color space data. It was calculated using (Equation (1)),

$$\Delta E = \sqrt{(L*_2 - L*_1)^2 + (a*_2 - a*_1)^2 + (b*_2 - b*_1)^2} \quad (1)$$

where, ΔE is the total difference between blank leather and leather/SeNPs, L^* is the lightness from black (0) to white (100), a^* is the red (+)/green (−) ratio and b^* is the yellow (+)/blue (−) ratio.

2.5.4. Exhaustion of SeNPs onto Leather

The UV/Vis spectrum of the SeNPs colloidal solutions before and after leather treatment was measured by UV/Vis spectrophotometer (Alpha-1860, Noble, IN, USA) to evaluate the SeNPs exhaustion.

2.5.5. Physical Properties of Leather

The color fastness of blank leather and leather/SeNPs was tested according to the following standard methods. They were determined using the AATCC (61-1996) [40], (8-1996) [41], and (16-2004) [42] tests for washing, rubbing, and light fastness, respectively. The tensile strength tests of blank leather and leather/SeNPs were performed according to ASTM D638-14 [43], using a tensile testing machine (Zwick Z010, Staufenberg, Germany). Additionally, the durability to washing was evaluated according to AATCC 61(2A)-1996 [44] after five washing cycles.

2.5.6. Cytotoxicity Test of Leather/SeNPs

Cytotoxicity was evaluated on human normal melanocyte cell line (HFB4) using MTT assay [45]. The leather/SeNPs dyed under the optimum conditions were sterilized and cut. Then it was plated on the six-well tissue culture plate. This plate was inoculated with cells and incubated for 24 h at 37 °C. The cell monolayer was washed twice using washing media after growth medium decantation. The cytotoxicity physical signs were checked in the tested cells. The tissue was picked up, and then 20 µL of 3-(4,5-dimethylthiazol-2-yl)-2,5-diphenyltetrazolium bromide dye (MTT) dissolved in phosphate buffer saline at a concentration of 5 mg/mL was added to each well with shaking for 5 min and incubated at 37 °C and 5% CO_2 for 5 h. The formazan was formed as MTT metabolite and resuspended in 200 µL dimethyl sulfoxide with shaking for 5 min. The absorbance was measured at 560 nm and the background was read at 620 nm, and then subtracted.

2.5.7. Antibacterial Activity

The antibacterial activity test for leather/SeNPs was conducted according to AATCC (147-2004) [46]. The bacterial strains were spread on the media using a sterile cotton swab. Antibacterial activity was evaluated against G+ve bacteria (*Bacillus cereus*) and G-ve bacteria (*Escherichia coli, Salmonella typhi* and *Pseudomonas aeruginosa*). Leather/SeNPs samples were cut to be square in shape and measuring 10 × 10 mm, while standard drugs such as tetracycline and ciprofloxacin were loaded in the disks to compare between their antibacterial activity and that of the tested leather/SeNPs. The petri dishes were then incubated at 37 °C for 24 h. while the growth inhibition zone diameters (mm) were determined.

2.5.8. Statistical Analysis

All tests have been performed by taking the average of three (samples) readings. The standard error of the mean was calculated according to the equation given below and found to be + (−) 0.1

$$SEX = S/\sqrt{n}$$

where S = sample standard deviation, and n = the number of observations of the sample

3. Results

3.1. Transmission Electron Microscopy (TEM) Analysis

The morphology and size of the synthesized SeNPs at different concentrations (25, 50 mM) were characterized using a transmission electron microscope. SeNPs prepared at the concentrations of 25 and 50 mM ranged from 36–77 nm and 41 to 149 nm, indicating the increase in average size with the increase in SeNPs concentration. The obtained TEM micrographs revealed that SeNPs were spherical in shape and well dispersed in the colloidal solution, especially that prepared at the low concentration. Furthermore, no agglomeration or deformation of SeNPs was observed, as displayed in Figure 1a,b. The width of bins in SeNPs histogram was 20 nm, as illustrated in Figure 1c, and these bins were centered at 40, 60, 80, 100, 120 and 140 nm. For example, all particles with diameters between 70 nm and 90 nm were considered together as particles with a diameter of 80 nm. In the case of low concentration (25 mM), SeNPs around 60 nm in size exhibited the highest percentage value of 76.06. On the other hand, the majority of SeNPs at the high concentration of 50 mM were from 70 to 110 nm, which confirmed the increase in SeNPs diameter and polydispersity as the increase in NPs concentration.

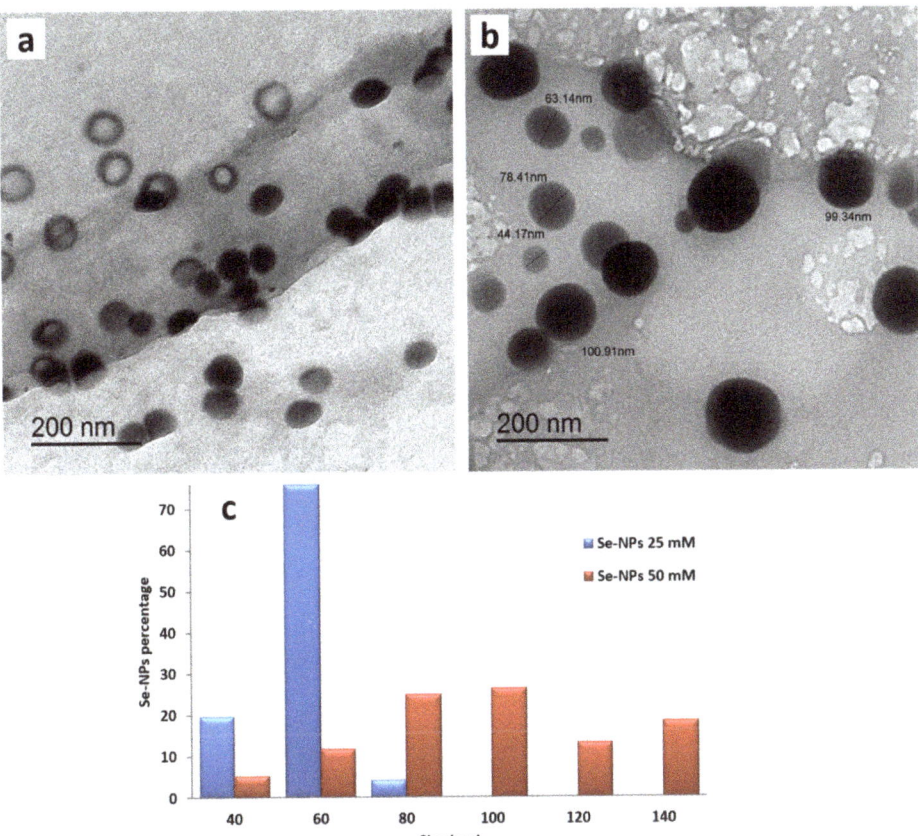

Figure 1. TEM images of SeNPs synthesized at different concentrations of (**a**) 25 mM and (**b**) 50 mM in addition to (**c**) the size distribution histogram of the prepared SeNPs.

Over and above, many SeNPs at the low concentration had a hollow shape (ring-shaped), while SeNPs at the concentration of 50 mM were ordinary solid spherical particles, illustrating that SeNPs specific surface area increased with the decrease in SeNPs concentration due to the reduction in NPs average size and the change in their morphology as shown in Figure 1.

3.2. Leather Characterization before and after Functionalization

3.2.1. Scanning Electron Microscopy (SEM) Analysis

The micrographs obtained from SEM analysis showed that the surface of untreated leather is typically clear with clean scales and smooth longitudinal collagen fibers as shown in Figure 2a. On the other hand, the SEM micrographs of leather/SeNPs revealed that the leather surface was coated sufficiently with a layer of SeNPs without aggregation and the SeNPs were distributed well on the leather surface as displayed in Figure 2b,c. Moreover, the chemical elements analysis of leather/SeNPs confirmed the presence of SeNPs on the surface after the dyeing process as shown in Figure 2d.

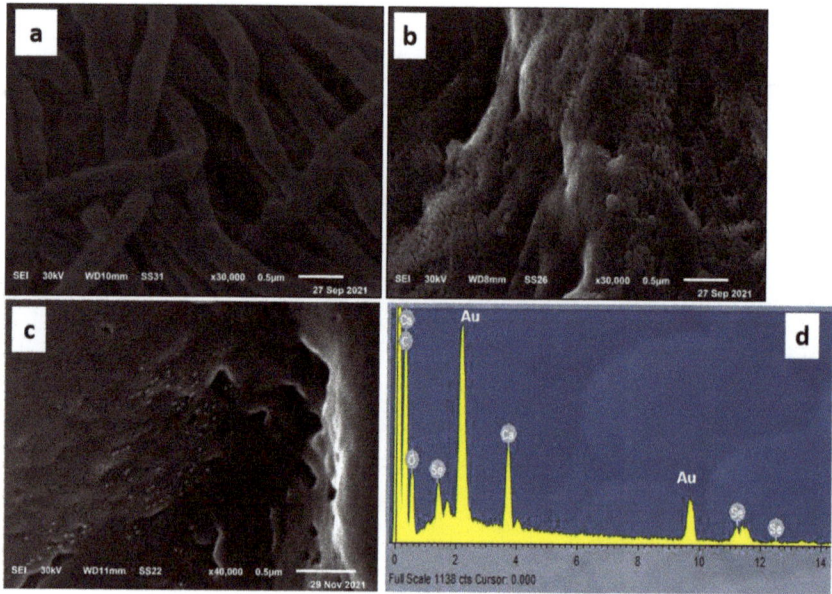

Figure 2. SEM micrographs of (**a**) blank leather and (**b**,**c**) leather dyed with SeNPs under the optimum conditions at different scales as well as (**d**) EDX spectrum of the dyed leather.

3.2.2. Raman Analysis

Raman analysis was conducted to confirm the deposition of SeNPs on the dyed leather surface as shown in Figure 3. The comparison between the spectra of leather surface before and after treatment with SeNPs was investigated in the range of 200–1600 cm^{-1}. After treatment two intense peaks at 230 and 295 cm^{-1} appeared as a result of SeNPs deposition. Similar results were obtained by Abou Elmaaty et al. [29] who treated the polypropylene fabric with SeNPs and found a peak around 236 cm^{-1} after SeNPs treatment, while Lukács et al. [47] found a pronounced peak at 292 cm^{-1} as an indicator for the presence of SeNPs that confirms the successful deposition of SeNPs on the treated leather surface.

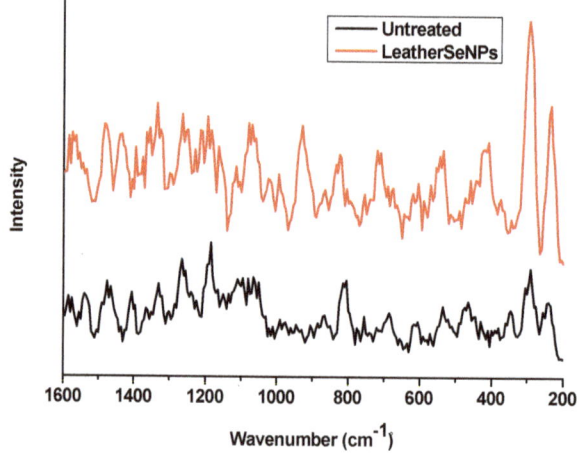

Figure 3. Raman spectra of untreated leather and leather/SeNPs$^-$.

3.2.3. Colorimetric Study

Table 1 showed the (L^*, a^*, b^*, C^*, h, K/S) values of blank leather and leather/SeNPs. The results revealed that the K/S value of leather/SeNPs was significantly higher than the blank leather. The (L^*, a^*, b^*) values for leather/SeNPs were varied considerably from blank leather. The color of blank leather is cream-white with a relatively high L^* value (74.9), low a^* value (8.6), and low b^* value (17.7). The lightness (L^*) value of leather/SeNPs was decreased due to the presence of SeNPs, resulting in the coloration of leather. The a^* values of leather/SeNPs was increased, the increment in a^* value indicated more redness color for the treated leather. The rise in b^* value of leather/SeNPs was observed as yellow/brownish, which confirming the coverage of leather with SeNPs. It can be concluded from the equation 1 and Table 1 that the total color difference (ΔE) between the two leathers is ~42.

Table 1. Color strength and colorimetric data of blank leather and leather/SeNPs.

Type	Sample	Color Parameters					
		L^*	a^*	b^*	C^*	h	K/S
Blank leather		74.9	8.6	17.7	22.5	64.0	1.7
Leather/SeNPs		37.1	26.1	24.8	19.6	43.5	18.7

Effect of Treatment pH on Color Strength (K/S)

To investigate the color changes of leather/SeNPs at different pH values under an ultrasonic water bath, Figure 4 showed the effect of pH on K/S of leather/SeNPs. The pH was adjusted from 3 to 8 to reflect the different colors expressed in leather/SeNPs. As shown, the maximum K/S value appeared at pH 6; this might be caused by overcrowding of SeNPs molecules into the leather matrix in this pH value due to the ultrasonic power [48]. The results indicated that the K/S of leather/SeNPs can be fine-tuned by controlling the pH. According to Lee and Kim et al. [49], the correlation between SeNPs and the leather can be attributed to pH of treatment bath [48].

Effect of Treatment Temperature on Color Strength (K/S)

Figure 5 demonstrated the K/S values of leathers/SeNPs at different temperatures. The results showed that the maximum K/S value was found to be 18.7 at 65 °C. In addition, when the temperature was increased above 65 °C, the K/S value was simultaneously decreased. This may be attributed to the tough muscle fiber of leather which did not allow the nanoparticles to enter the leather matrix, causing a lower SeNPs uptake. Moreover, high temperatures can cause the leather to shrink, harden, stiffen, and become brittle. It can be concluded that the temperature of 65 °C was set to be the optimum temperature for binding SeNPs into the leather [48].

Effect of Treatment Time on Color Strength (K/S)

Figure 6 exhibited the effect of treatment time on SeNPs uptake by leather. As depicted, the K/S was increased upon increasing time of treatment. The optimum K/S value of 18.7 was achieved at 60 min after which, K/S started to decrease with increasing time. The decrease in K/S as the increase in time from 90 to 120 min may be due to the aggregation of SeNPs as a result of SeNPs accumulation upon the leather surface [48].

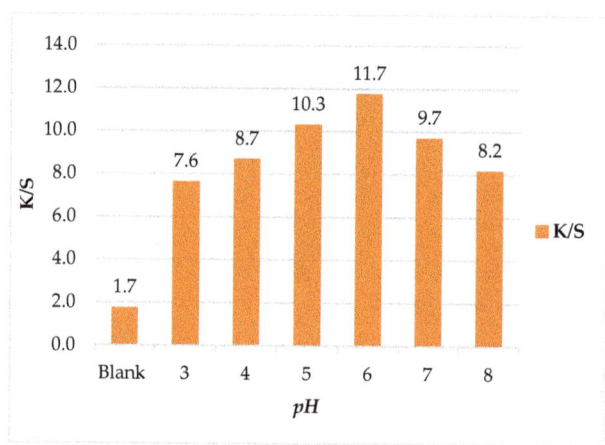

Figure 4. Effect of *pH* on color strength (*K/S*) of leather/SeNPs.

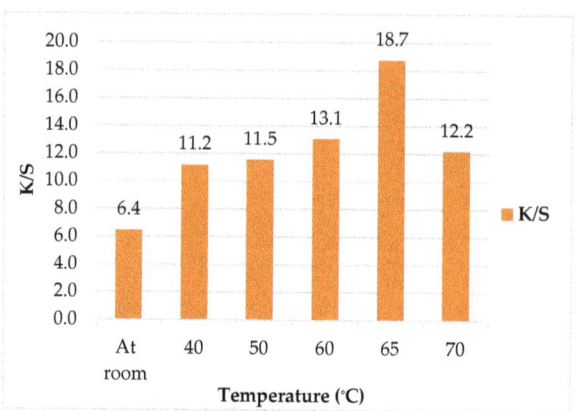

Figure 5. Effect of treatment temperature on color strength (*K/S*) of leather/SeNPs.

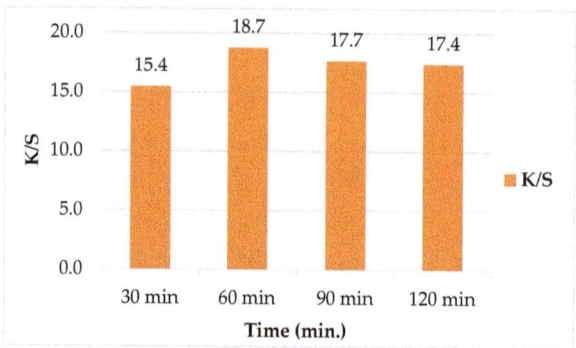

Figure 6. Effect of treatment time on color strength (*K/S*) of leather/SeNPs.

Effect of SeNPs Concentration on Color Strength (*K/S*)

The color strength of the leather/SeNPs was found to be dependent on SeNPs concentration. Upon increasing the concentration of SeNPs, the *K/S* value was decreased as shown in Figure 7. The highest *K/S* value for the treated leather was monitored at the lowest concentration. This proved that treating leather with lower concentration of SeNPs (25 mM) was adequate to achieve the optimum *K/S*. The aforementioned behavior could be attributed to the thermal stability of the leather, as it increased at low SeNPs concentration and decreased at high SeNPs concentration. The improvement in thermal stability could be recognized as the crosslinking between SeNPs and collagen of leather [4].

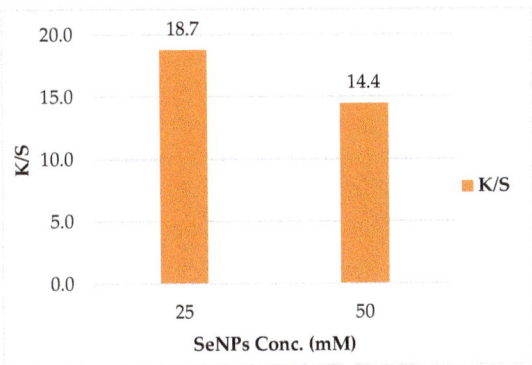

Figure 7. Effect of SeNPs concentrations on color strength (*K/S*) of leather/SeNPs.

3.2.4. Exhaustion of SeNPs onto Leather

The UV/Visible spectrum of SeNPs colloidal solution was measured before and after leather treatment to evaluate the SeNPs exhaustion by treated leather as shown in Figure 8.

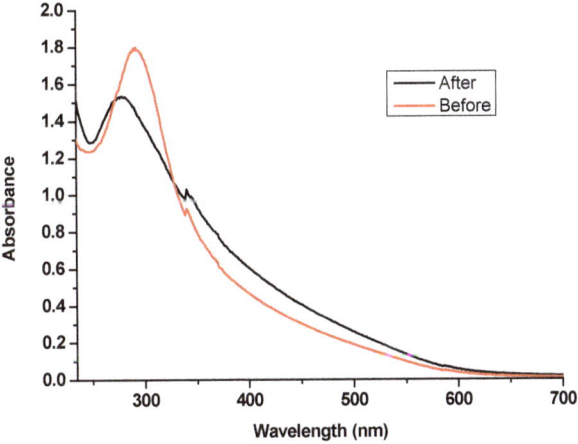

Figure 8. UV/Visible spectrum of SeNPs before and after leather treatment.

The absorbance of SeNPs was measured in the range of 235 to 700 nm. The absorption spectrum of SeNPs showed a sharp absorption band at 290 nm before the exhaustion, while SeNPs spectrum after leather treatment decreased obviously after exhaustion, confirming the deposition of SeNPs on the treated leather.

3.2.5. Physical Properties of Leather/SeNPs

The tensile modulus and elongation at break % values were listed in Table 2. The obtained results indicated that there were no significant differences about the mechanical properties of leathers, mainly between the leather/SeNPs compared to the blank leather.

The color fastness properties of leather/SeNPs were evaluated and the results were tabulated in Table 2. Leather/SeNPs revealed excellent fastness results referring to the chemical stability of the SeNPs onto the leather surfaces along with long-term durable interactions between the SeNPs and the leather surface [4].

Table 2. Properties of the Leather and Leather/SeNPs under optimum conditions.

Sample	Wash Fastness		Rubbing Fastness		Light Fastness	Tensile Strength	
	St.	Alt.	Dry	Wet		Tensile Modulus, [a] MPa	Elongation, [b] %
Blank leather	-	-	-	-	-	8	36
Leather covered with SeNPs	5	5	5	4/5	4/5	7	32
Leather covered with SeNPs after 5 washing cycles (durability test)	5	5	5	4/5	4/5	7	32

Treatment conditions: SeNPs, (25 mM); Time (60) min; Temp. (65) °C; pH 6. [a] Standard deviation for untreated leather = 0.5, standard deviation for treated leather = 0.5. [b] Standard deviation for untreated leather = 0.5, standard deviation for treated leather = 0.5.

3.2.6. Cytotoxicity of Leather/SeNPs

Leather/SeNPs cytotoxicity was evaluated against healthy human melanocyte cell line (HFB4) using MTT assay. The viability of cells in the presence of leather/SeNPs was 90.67% compared to that of negative control. The average relative cell viability is over 70% [50], indicating the low toxicity of leather/SeNPs toward human skin.

3.2.7. Antibacterial Activity

The antibacterial activity of leather/SeNPs was evaluated against four bacterial strains, including *Bacillus cereus* as Gram-positive bacteria in addition to *Escherichia coli*, *Pseudomonas aeruginosa* and *salmonella typhi* as Gram-negative bacteria. The results revealed that the leather/SeNPs showed outstanding antibacterial activity against the tested strains in comparison with standard drugs as listed in Table 3. Leather treated with 25 mM SeNPs colloidal solution was more effective against *Escherichia coli*, *Pseudomonas aeruginosa* and *Bacillus cereus* than that treated with high concentration (50 mM) due to the difference in average size. SeNPs (25 mM) with small average size exhibited higher specific area and in contact with bacterial cells more than that at the high concentration (50 mM) [51]. There was no obvious variation in inhibition zone diameters for leather/SeNPs before and after five washing cycles that confirms the durability of the leather/SeNPs samples at the tested concentrations. Moreover, the cross-linking between leather and SeNPs as antimicrobial agents protects leather against laundering and mechanical abrasion that makes the leather products more durable [22].

Table 3. The inhibition zone (mm) values of leathers/SeNPs treated with different SeNPs concentrations.

Substrate	Bacillus cereus (G+)	Escherichia coli (G−)	Pseudomonas aeruginosa (G−)	Salmonella typhi (G−)
Leather/SeNPs (25 mM)	20	15	19	16
Leather/SeNPs (50 mM)	18	9	15	18
Leather/SeNPs (25 mM); after 5 washing cycles	20	14	18	14
Leather/SeNPs (50 mM); after 5 washing cycles	17	8	15	18
Tetracycline (30 µg)	15	19	16	13
Ciprofloxacin (10 µg)	18	21	17	15

The antibacterial effect on bacteria may be due to the release of ions or the formation of reactive oxygen species that lead to DNA damage. Furthermore, the deposited SeNPs can contact bacteria or fungi cells much easier than colloidal form [22].

4. Conclusions

Leather material was successfully colored and functionalized utilizing selenium nanoparticles (SeNPs) which were synthesized and deposited simultaneously onto the leather surface. The SeNPs decorated the leather surface with shining colors, which can be controlled by adjusting the *pH* at 6, treatment time for 60 min., treatment temperature at 65 °C and SeNPs concentration of 25 mM. The results showed that a yellow/brown color was imparted to the leather after the implementation with SeNPs by ultrasound technique. Moreover, the colored leather samples acquired good color fastness to rub, wash, and light. Leathers/SeNPs exhibited excellent antibacterial activities against models of bacteria, including Gram-positive bacteria (*Bacillus cereus*) and Gram-negative bacteria (*Pseudomonas aeruginosa, Escherichia coli* and *Salmonella typhi*). The results of coloration, cytotoxicity and antibacterial properties clarified that the SeNPs can be used to impart color and antibacterial properties to leather material. The proposed methodology emphasized the effectiveness and applicability of this simple approach to the footwear industry to color the leather as well as prevent the spread of bacterial infection promoted by humidity, poor breathability and temperature.

Author Contributions: T.A.E. conceived the original idea of this study and participated in design and coordination as well as manuscript drafting. S.A.A. participated in the manuscript writing and interpreted the obtained results from colorimetric and mechanical properties study. K.S.-A. prepared the SeNPs and interpreted the obtained results obtained from TEM, SEM analysis, antimicrobial and cytotoxicity tests as well as participation in manuscript writing. K.E.-K. was responsible for the dyeing experiments and leather finishing steps in addition to participation in SeNPs preparation. R.M.A. provided the tanned leather required for this study. All authors have read and agreed to the published version of the manuscript.

Funding: This paper is based upon work supported by Science, Technology & Innovation Funding Authority (STDF); Egypt under grant (POST GRADUATE SUPPORT GRANT, CALL-1).

Institutional Review Board Statement: Not applicable.

Informed Consent Statement: Not applicable.

Data Availability Statement: The data presented in this study are available on request from the corresponding author.

Conflicts of Interest: The authors declare no conflict of interest.

References

1. Velmurugan, P.; Shim, J.; Bang, K.-S.; Oh, B.-T. Gold nanoparticles mediated coloring of fabrics and leather for antibacterial activity. *J. Photochem. Photobiol. B Biol.* **2016**, *160*, 102–109. [CrossRef]
2. Hu, J.; Deng, W. Application of supercritical carbon dioxide for leather processing. *J. Clean. Prod.* **2016**, *113*, 931–946. [CrossRef]
3. Su, Y.; Li, P.; Gao, D.; Lyu, B.; Ma, J.; Zhang, J.; Lyu, L. High-efficiency antibacterial and anti-mildew properties under self-assembly: An environmentally friendly nanocomposite. *Adv. Powder Technol.* **2021**, *32*, 2433–2440. [CrossRef]
4. Elsayed, H.; Hasanin, M.; Rehan, M. Enhancement of multifunctional properties of leather surface decorated with silver nanoparticles (Ag NPs). *J. Mol. Struct.* **2021**, *1234*, 130130. [CrossRef]
5. Sportelli, M.C.; Picca, R.A.; Paladini, F.; Mangone, A.; Giannossa, L.C.; Franco, C.D.; Gallo, A.I.; Valentini, A.; Sannino, A.; Pollini, M.; et al. Spectroscopic Characterization and Nanosafety of Ag-Modified Antibacterial Leather and Leatherette. *Nanomaterials* **2017**, *7*, 203. [CrossRef] [PubMed]
6. Gurera, D.; Bhushan, B. Fabrication of bioinspired superliquiphobic synthetic leather with self-cleaning and low adhesion. *Colloids Surf. A Physicochem. Eng. Asp.* **2018**, *545*, 130–137. [CrossRef]
7. Bao, Y.; Feng, C.; Wang, C.; Ma, J.; Tian, C. Hygienic, antibacterial, UV-shielding performance of polyacrylate/ZnO composite coatings on a leather matrix. *Colloids Surf. A Physicochem. Eng. Asp.* **2017**, *518*, 232–240. [CrossRef]

8. Khan, S.A.; Shahid, S.; Kanwal, S.; Rizwan, K.; Mahmood, T.; Ayub, K. Synthesis of novel metal complexes of 2-((phenyl (2-(4-sulfophenyl) hydrazono) methyl) diazenyl) benzoic acid formazan dyes: Characterization, antimicrobial and optical properties studies on leather. *J. Mol. Struct.* **2019**, *1175*, 73–89. [CrossRef]
9. Liu, G.; Li, K.; Luo, Q.; Wang, H.; Zhang, Z. PEGylated chitosan protected silver nanoparticles as water-borne coating for leather with antibacterial property. *J. Colloid Interface Sci.* **2017**, *490*, 642–651. [CrossRef] [PubMed]
10. Tamil Selvi, A.; Aravindhan, R.; Madhan, B.; Raghava Rao, J. Studies on the application of natural dye extract from Bixa orellana seeds for dyeing and finishing of leather. *Ind. Crop. Prod.* **2013**, *43*, 84–86. [CrossRef]
11. Gaidau, C.; Ignat, M.; Iordache, O.; Madalina, L.; Piticescu, R.; Ditu, L.-M.; Ionescu, M. ZnO Nanoparticles for Antimicrobial Treatment of Leather Surface. *Revista De Chimie* **2018**, *69*, 767–771. [CrossRef]
12. Erciyes, A.; Ocak, B. Physico-mechanical, thermal, and ultraviolet light barrier properties of collagen hydrolysate films from leather solid wastes incorporated with nano TiO_2. *Polym. Compos.* **2019**, *40*, 4716–4725. [CrossRef]
13. Solangi, B.; Nawaz, H.; Solangi, B.; Zehra, B.; Nadeem, U. Preparation of Nano Zinc Oxide and its Application in Leather as a Retanning and Antibacterial Agent Preparation of Nano Zinc Oxide and its Application in Leather as a Retanning and Antibacterial Agent. *Can. J. Sci. Ind. Res.* **2011**, *2*, 164–170.
14. Liu, J.; Ma, J.; Bao, Y.; Wang, J.; Tang, H.; Zhang, L. Polyacrylate/Surface-Modified ZnO Nanocomposite as Film-Forming Agent for Leather Finishing. *Int. J. Polym. Mater. Polym. Biomater.* **2014**, *63*, 809–814. [CrossRef]
15. Nidhin, D.M.; Rathinam, A.; Sreeram, K.J. GREEN synthesis of monodispersed iron oxide nanoparticles for leather finishing. *J. Am. Leather Chem. Assoc.* **2014**, *109*, 184–188.
16. Chen, Y.; Fan, H.; Shi, B. Nanotechnologies for leather manufacturing: A review. *J. Am. Leather Chem. Assoc.* **2011**, *106*, 260–273.
17. Yorgancioglu, A.; Bayramoglu, E.; Renner, M. Preparation of Antibacterial Fatliquoring Agents Containing Zinc Oxide Nanoparticles for Leather Industry. *J. Am. Leather Chem. Assoc.* **2019**, *114*, 171–179.
18. Petica, A.; Gaidau, C.; Ignat, M.; Sendrea, C.; Anicai, L. Doped TiO_2 nanophotocatalysts for leather surface finishing with self-cleaning properties. *J. Coat. Technol. Res.* **2015**, *12*, 1153–1163. [CrossRef]
19. Ma, J.; Zhang, X.; Bao, Y.; Liu, J. A facile spraying method for fabricating superhydrophobic leather coating. *Colloids Surf. A Physicochem. Eng. Asp.* **2015**, *472*, 21–25. [CrossRef]
20. Velmurugan, P.; Lee, S.-M.; Cho, M.; Park, J.-H.; Seo, S.-K.; Myung, H.; Bang, K.-S.; Oh, B.-T. Antibacterial activity of silver nanoparticle-coated fabric and leather against odor and skin infection causing bacteria. *Appl. Microbiol. Biotechnol.* **2014**, *98*, 8179–8189. [CrossRef] [PubMed]
21. Velmurugan, P.; Cho, M.; Lee, S.-M.; Park, J.-H.; Bae, S.; Oh, B.-T. Antimicrobial fabrication of cotton fabric and leather using green-synthesized nanosilver. *Carbohydr. Polym.* **2014**, *106*, 319–325. [CrossRef] [PubMed]
22. Xiang, J.; Ma, L.; Su, H.; Xiong, J.; Li, K.; Xia, Q.; Liu, G. Layer-by-layer assembly of antibacterial composite coating for leather with cross-link enhanced durability against laundry and abrasion. *Appl. Surf. Sci.* **2018**, *458*, 978–987. [CrossRef]
23. Zohreh, M.; Ani, I.; Peiman, V. Antibacterial improvement of leather by surface modification using corona discharge and silver nanoparticles application. *J. Sci. Technol.* **2014**, *5*, 1–15.
24. Elsayed, H.; Attia, R.; Mohamed, O.; Haroun, A.; El-Sayed, N. Preparation of Polyurethane Silicon Oxide Nanomaterials as a Binder in Leather Finishing. *Fibers Polym.* **2018**, *19*, 832–842. [CrossRef]
25. Mohamed, O.; Elsayed, H.; Attia, R.; Haroun, A.; El-Sayed, N. Preparation of acrylic silicon dioxide nanoparticles as a binder for leather finishing. *Adv. Polym. Technol.* **2018**, *37*, 3276–3286. [CrossRef]
26. Carvalho, I.; Ferdov, S.; Mansilla, C.; Marques, S.M.; Cerqueira, M.A.; Pastrana, L.M.; Henriques, M.; Gaidau, C.; Ferreira, P.; Carvalho, S. Development of antimicrobial leather modified with Ag–TiO_2 nanoparticles for footwear industry. *Sci. Technol. Mater.* **2018**, *30*, 60–68. [CrossRef]
27. Gaidau, C.; Petica, A.; Ignat, M.; Iordache, O.; Ditu, L.-M.; Ionescu, M. Enhanced photocatalysts based on Ag-TiO_2 and Ag-N-TiO_2 nanoparticles for multifunctional leather surface coating. *Open Chem.* **2016**, *14*, 383–392. [CrossRef]
28. Kaygusuz, M.; Meyer, M.; Aslan, A. The Effect of TiO_2-SiO_2 Nanocomposite on the Performance Characteristics of Leather. *Mater. Res. Ibero-Am. J. Mater.* **2017**, *20*, 1103–1110. [CrossRef]
29. AbouElmaaty, T.; Abdeldayem, S.A.; Ramadan, S.M.; Sayed-Ahmed, K.; Plutino, M.R. Coloration and Multi-Functionalization of Polypropylene Fabrics with Selenium Nanoparticles. *Polymers* **2021**, *13*, 2483. [CrossRef]
30. Abou Elmaaty, T.M.; Abdeldayem, S.A.; Elshafai, N. Simultaneous Thermochromic Pigment Printing and Se-NP Multifunctional Finishing of Cotton Fabrics for Smart Childrenswear. *Cloth. Text. Res. J.* **2020**, *38*, 182–195. [CrossRef]
31. Razmkhah, M.; Montazer, M.; Bashiri Rezaie, A.; Rad, M.M. Facile technique for wool coloration via locally forming of nano selenium photocatalyst imparting antibacterial and UV protection properties. *J. Ind. Eng. Chem.* **2021**, *101*, 153–164. [CrossRef]
32. Elmaaty, T.M.A.; Raouf, S.; Sayed-Ahmed, K. Novel One Step Printing and Functional Finishing of Wool Fabric Using Selenium Nanoparticles. *Fibers Polym.* **2020**, *21*, 1983–1991. [CrossRef]
33. Jia, X.; Liu, Q.; Zou, S.; Xu, X.; Zhang, L. Construction of selenium nanoparticles/β-glucan composites for enhancement of the antitumor activity. *Carbohydr. Polym.* **2015**, *117*, 434–442. [CrossRef]
34. Chen, W.; Li, Y.; Yang, S.; Yue, L.; Jiang, Q.; Xia, W. Synthesis and antioxidant properties of chitosan and carboxymethyl chitosan-stabilized selenium nanoparticles. *Carbohydr. Polym.* **2015**, *132*, 574–581. [CrossRef]
35. Kong, H.; Yang, J.; Zhang, Y.; Fang, Y.; Nishinari, K.; Phillips, G.O. Synthesis and antioxidant properties of gum arabic-stabilized selenium nanoparticles. *Int. J. Biol. Macromol.* **2014**, *65*, 155–162. [CrossRef]

36. Wang, Q.; Barnes, L.-M.; Maslakov, K.I.; Howell, C.A.; Illsley, M.J.; Dyer, P.; Savina, I.N. In situ synthesis of silver or selenium nanoparticles on cationized cellulose fabrics for antimicrobial application. *Mater. Sci. Eng. C* **2021**, *121*, 111859. [CrossRef]
37. Shakibaie, M.; Forootanfar, H.; Golkari, Y.; Mohammadi-Khorsand, T.; Shakibaie, M.R. Anti-biofilm activity of biogenic selenium nanoparticles and selenium dioxide against clinical isolates of Staphylococcus aureus, Pseudomonas aeruginosa, and Proteus mirabilis. *J. Trace Elem. Med. Biol.* **2015**, *29*, 235–241. [CrossRef]
38. Biswas, D.P.; O'Brien-Simpson, N.M.; Reynolds, E.C.; O'Connor, A.J.; Tran, P.A. Comparative study of novel in situ decorated porous chitosan-selenium scaffolds and porous chitosan-silver scaffolds towards antimicrobial wound dressing application. *J. Colloid Interface Sci.* **2018**, *515*, 78–91. [CrossRef] [PubMed]
39. Nagar, R. Replacement of Lime with Sodium Hydroxide in Leather Tanning. *Environ. Sci.* **2020**, *15*, 159–172.
40. American Association of Textile Chemists and Colorists. *AATTCC 61-1996: Colorfastness to Laundering: Accelerated*; American Association of Textile Chemists and Colorists: Research Triangle Park, NC, USA, 1996.
41. American Association of Textile Chemists and Colorists. *AATCC Test Method 8-1996, Colorfastness to Crocking*; American Association of Textile Chemists and Colorists: Research Triangle Park, NC, USA, 1996.
42. American Association of Textile Chemists and Colorists. *AATCC16-2004: Colorfastness to Light*; American Association of Textile Chemists and Colorists: Research Triangle Park, NC, USA, 2004.
43. ASTM International. *ASTM D638-14, Standard Test Method for Tensile Properties of Plastics*; ASTM International: West Conshohochen, PA, USA, 2015.
44. American Association of Textile Chemists and Colorists. *AATCC 61(2A): Colorfastness to Laundering*; American Association of Textile Chemists and Colorists: Research Triangle Park, NC, USA, 1996.
45. Mosmann, T. Rapid colorimetric assay for cellular growth and survival: Application to proliferation and cytotoxicity assays. *J. Immunol. Methods* **1983**, *65*, 55–63. [CrossRef]
46. American Association of Textile Chemists and Colorists. *AATTCC 147-2004: Antibacterial Activity Assessment of Textile Materials*; American Association of Textile Chemists and Colorists: Research Triangle Park, NC, USA, 2010.
47. Lukács, R.; Veres, M.; Shimakawa, K.; Kugler, S. On photoinduced volume change in amorphous selenium: Quantum chemical calculation and Raman spectroscopy. *J. Appl. Phys.* **2010**, *107*, 073517. [CrossRef]
48. Velmurugan, P.; Vedhanayakisri, K.A.; Park, Y.-J.; Jin, J.-S.; Oh, B.-T. Use of Aronia melanocarpa Fruit Dye Combined with Silver Nanoparticles to Dye Fabrics and Leather and Assessment of Its Antibacterial Potential Against Skin Bacteria. *Fibers Polym.* **2019**, *20*, 302–311. [CrossRef]
49. Lee, Y.-H.; Kim, H.-D. Dyeing properties and colour fastness of cotton and silk fabrics dyed with *Cassia tora* L. extract. *Fibers Polym.* **2004**, *5*, 303–308. [CrossRef]
50. Kangwansupamonkon, W.; Lauruengtana, V.; Surassmo, S.; Ruktanonchai, U. Antibacterial effect of apatite-coated titanium dioxide for textiles applications. *Nanomed. Nanotechnol. Biol. Med.* **2009**, *5*, 240–249. [CrossRef]
51. Inam, M.; Foster, J.C.; Gao, J.; Hong, Y.; Du, J.; Dove, A.P.; O'Reilly, R.K. Size and shape affects the antimicrobial activity of quaternized nanoparticles. *J. Polym. Sci. Part A Polym. Chem.* **2019**, *57*, 255–259. [CrossRef]

MDPI
St. Alban-Anlage 66
4052 Basel
Switzerland
Tel. +41 61 683 77 34
Fax +41 61 302 89 18
www.mdpi.com

Polymers Editorial Office
E-mail: polymers@mdpi.com
www.mdpi.com/journal/polymers

www.ingramcontent.com/pod-product-compliance
Lightning Source LLC
LaVergne TN
LVHW070652100526
838202LV00013B/944